The Thirteenth-Century Animal Turn

Nigel Harris

The Thirteenth-Century Animal Turn

Medieval and Twenty-First-Century Perspectives

palgrave
macmillan

Nigel Harris
University of Birmingham
Birmingham, UK

ISBN 978-3-030-50660-5   ISBN 978-3-030-50661-2   (eBook)
https://doi.org/10.1007/978-3-030-50661-2

Cover pattern © Melisa Hasan

This Palgrave Macmillan imprint is published by the registered company Springer Nature Switzerland AG.
The registered company address is: Gewerbestrasse 11, 6330 Cham, Switzerland

To Katharine, Elizabeth, David, Alfie and the cats
And to the memory of James Stuart Pascoe (1932–2015): "A Man
for All Seasons"

PREFACE

This book has not been long in the writing, but has probably been in a state of sub- or semi-conscious gestation ever since I began my post-graduate studies in 1981. Thanks to the rigorous but always supportive training provided by Nigel Palmer, these prepared me for a research career which has been punctuated by various publications about animals and their spiritual interpretation in the Middle Ages; and in recent years my thinking on the subject has been stimulated and challenged by several books which have emanated from the relatively new field of (Human-) Animal Studies. Meanwhile the question as to my own personal relationship with the natural world has become more urgent, indeed existential in recent years, thanks to my having inherited a farm in Cornwall in 2015.

It is, then, only right that I should dedicate this volume to the memory of the remarkable Renaissance man and animal-lover who made this gift – as well as, of course, to the human and non-human members of my more immediate family, to whom I owe so much at all times. Acknowledgement must also be made of Andrew Ginger, whose brainwave it was that my ideas in their current form would fit well with the length and ethos of a Palgrave Pivot volume; of my commissioning editor Allie Troyanos, for believing in the project; and of Palgrave's two anonymous readers, whose comments on my first draft – full and sympathetic, but also bristling with constructive insights – were a model of their kind. All errors or other inadequacies are, of course, my own – as indeed (unless acknowledgement to the contrary is made) are the book's many translations into English.

Hall Green, Birmingham   Nigel Harris
April 2020

CONTENTS

CONTENTS

CHAPTER 1

Introduction

Abstract This introductory chapter defines the concept of a cultural turn. Such phenomena begin with a heightened scholarly awareness of a particular aspect of reality – an awareness which, however, proves over time to be intellectually and methodologically productive and to have a significant impact on the world beyond the academy. On this basis the chapter argues that an Animal Turn took place in the thirteenth century which is comparable to the one currently in progress in the twenty-first. It goes on to situate the present volume within current research in medieval (Human-) Animal Studies and the German tradition of studying allegorical interpretations of animals; and it defines its temporal and geographical terms of reference.

Keywords Animal Turn • Medieval • Thirteenth century

I

The title of this book raises several questions. What is an "Animal Turn"? What, if anything, might it have in common with, say, an interpretative, iconic, performative, reflexive, post-colonial, translational or spatial turn? And, not least, can one defensibly speak of such cultural turns in respect of any centuries before the late twentieth?

© The Author(s) 2020
N. Harris, *The Thirteenth-Century Animal Turn,*
https://doi.org/10.1007/978-3-030-50661-2_1

There is, as far as I know, no agreed definition of what a "turn" in this sense is – so we must begin by attempting one that is relevant at least to the concerns of this volume. Turns are not, I think, quite the same as paradigm shifts (*pace* Kompatscher 18). The latter are more fundamental and hence rarer developments which have the function of replacing one set of values or perspectives by another. Turns, however, do not – usually – replace existing paradigms, but rather add to, complement and above all challenge them.

At its most basic level, a turn is characterized by a heightened scholarly awareness of a particular subject or approach to that subject. Karl Schlögel (265) puts this well, if with rather too many distancing adverbs:

> The turn is apparently the modern way of referring to the heightened aware-ness of dimensions and aspects that were previously neglected... It suggests that a multitude of very different perspectives are possible on the same sub-ject. It is apparently an enrichment of the act of seeing, perceiving and pro-cessing. Turns ... are evidently an indication that something is afoot: an opening, an expansion, a pluralization of dimensions (quoted from Bachmann-Medick, 15).

As an example of what this might mean specifically in the case of literary animals, Roland Borgards (2016, vii) alludes to the fleas on the collar of the gatekeeper in Franz Kafka's famously impenetrable story *Vor dem Gesetz/Before the Law*. These are almost invariably overlooked by scholars, and presumably also by other readers. Yet they are an intrinsic part of the story and, when properly considered, can add an important extra dimen-sion to its interpretation. Especially given that the "man from the coun-try" only becomes aware of them after several years of looking with increasing frustration at his interlocutor (the gatekeeper), they remind us of the simultaneous ubiquity and near-invisibility of many animals in a human-dominated world; and they also reveal much about the character and predicament of the "man from the country" himself – especially when he proceeds to speak to the fleas and ask them for their support in pleading with their host. Kafka's narrator describes this behaviour as "childish"; but it is also an extremely powerful image of existential desolation and despair.

A turn is not a turn, however, if it only affects academic readers of Kafka. It has to be more than a heightened scholarly awareness, more than something conjured up in the seclusion of some ivory tower. Rather, it has to interact with and, to some extent, channel concerns that are felt also beyond the narrow confines of academia. By this criterion, the recent surge in scholarly interest in animals certainly bears the hallmarks of a genuine turn, rather than a professorial fad: it is intimately connected with things that are "afoot" in many Western societies more generally. Linda Kalof's summary of these factors (2017, 1) is far from complete, but brilliantly economical:

> The remarkable flourishing of animal studies is due to the widespread recognition of (1) the commodification of animals in a wide variety of human contexts such as the use of animals as food, labor, and the objects of spectacle and science; (2) the degradation of the natural world, a staggering loss of animal habitat, and species extinction; and (3) our increasing need to coexist with other animals in urban, rural, and natural contexts.

Finally, for a scholarly vogue to merit description as a turn, it must make not just a quantitative, but also a qualitative difference to intellectual life. In other words, it must not just result in the production of more books and articles, but must involve or occasion new or revised methodologies that genuinely "enrich the act of seeing". Again this is something that the current Animal Turn has done with marked success. It has encouraged new levels of inter- and crossdisciplinarity, not merely within the Humanities but also in bridging the historically stifling divides between "arts", "sciences" and "social sciences". Moreover it has introduced new theorists and theoretical perspectives into academic discussions of animals: Jacques Derrida, Claude Lévi-Strauss and Donna Haraway, to name but three, have had an enormous impact on many scholars working on animals in the last twenty years, and hence an at least indirect influence on the present volume. Above all, perhaps, Animal Studies has consistently integrated into its analysis the perspectives, subjectivity and agency of animals themselves – be they domesticated or wild, real or imaginary. Kafka's fleas are just one example among many.

II

The present volume responds to the current Animal Turn by considering two questions that I, as a twenty-first-century medievalist, have been asking myself for some time: is it possible to speak of an Animal Turn (understood, *mutatis mutandis*, as outlined above) in respect of any medieval century? And, if so, which one? The answers to be proposed in what follows are – with of course all the usual scholarly caveats – "yes", and "the thirteenth".

When we contemplate the thirteenth century in the light of the three characteristics of a turn we have just adumbrated, it seems legitimate to speak of it having accommodated an Animal Turn. First, there can be no doubt that, especially from around 1220 onwards, thirteenth-century scholars had a heightened awareness of animals. This came from two main sources. On the one hand, Michael Scot's Latin translation of the natural historical writings of Aristotle had a huge impact on thirteenth-century academic life, visible not least in the enormous upsurge of learned writing about animals in Latin. Works as diverse as the encyclopaedias of Albert the Great (Albertus Magnus), Thomas of Cantimpré and Bartholomaeus Anglicus, the collections of nature exempla that arose in their wake, and Emperor Frederick II's monumental treatise on falconry would have been inconceivable without the impetus to study animals provided by Aristotle. At roughly the same time that Michel Scot was translating Aristotle, however, Pope Innocent III was instituting reforms at the Fourth Lateran Council of the Church that in short order would give rise, over much of Europe, to a significant increase in the amount of sermons preached and moral instruction delivered in parish churches. Given that stories involving animals were a highly popular feature of such teaching designed for the laity, this too led to a major growth in moral theological literature about them.

The latter development was one way in which the heightened scholarly awareness of animals was grounded outside the usual circles of learned Latinity – whose functions in medieval society can be equated at least roughly with those of the university-based academic community of today. Still more striking evidence of an Animal Turn that had an impact beyond studious intellectuals can be found, however, in the chivalric world of the lay aristocracy. This too witnessed a variety of changing perspectives and practices with regard to animals. For example, technological and tactical developments changed certain aspects of a knight's partnership with his warhorse; and a perceptible boom in the use of animals in heraldic insignia both reflected and stimulated an enhanced interest in animals more generally – including mythical ones and those not indigenous to Europe. Underlying this development seems to have been a greater degree of awareness – very noticeable in imaginative literature – of animals' usefulness for the construction or reconstruction of chivalric identity in an age where it was commonly seen as being under threat. Moreover the widely documented association between aristocratic ladies and lapdogs, which derived from an exponential growth in the keeping of pets from the early thirteenth century, implies that such considerations to some extent transcended gender boundaries.

Our third defining characteristic of a turn was that it should involve a new methodology, a new way of seeing; and this certainly obtained in respect of the thirteenth-century Animal Turn we are postulating. Again the key figure was Aristotle, as channelled by Michael Scot. In contrast to the traditional, Augustinian view that the study of animals was in essence a means to an end (that of teaching humans about God and how they might relate to him), Aristotle demonstrated, and encouraged in others, a strong interest in animals in their own right and for their own sake. In doing so he modelled a new reliance on observation and reason, to supplement data passed down by venerable authorities; and he – and the encyclopaedists who followed him – evinced a strong desire to systematize what was known about the natural world, in a way that seems to us now in some respects more modern than medieval.

The last paragraph in particular could easily give the impression that we are dealing here with a fundamental paradigm shift, rather than a modifying, challenging cultural turn. That, however, is not really the case. The thirteenth century was, not only but not least in relation to animals, an age of continuity as well as of innovation. This can be seen in the significant levels of physical violence, often casual and brutal, which thirteenth-century people continued to visit upon animals on a regular basis. Moreover the new Aristotelian methods of scientific analysis did not entirely supersede the older, Augustinian idea of nature as a book designed to teach us about God. Indeed in many ways they complemented, enhanced and revivified the inherited didactic traditions, not least by anchoring them more firmly in the reality of what animals were actually like.

Most fundamentally perhaps, the pre-Darwinian idea of a clear anthropological difference separating rational humans from irrational animals remained intact in 1300, just as it had in 1200. Self-evidently no medieval author would think of her- or himself as an animal – as I, and no doubt most people today, are happy to do. Nevertheless various aspects of thirteenth-century culture did pose informed and challenging questions about the nature of animality – and hence, inevitably, about the nature of humanity as well; and it was becoming increasingly clear that these questions could no longer be fobbed off with straightforward rehearsals of time-honoured notions about human uniqueness and superiority.

III

In what follows I argue the case for positing a thirteenth-century Animal Turn in several stages. Chapter 2 focuses on Aristotle and his influence – with regard both to the information his newly translated works transmitted, and to the intellectual methods and habits they fostered. Two paradigmatically Aristotelian Latin works from the thirteenth century feature particularly prominently in the chapter, namely the *De animalibus* of Albert the Great (part commentary on Aristotle and part nature encyclopaedia), and the *De arte venandi cum avibus* of Emperor Frederick II. I also consider Aristotle's complex and sometimes confused musings about the relationship between the human and the animal, and trace their influence on sources as diverse as Albert's conception of pygmies and comic tales which portray Aristotle himself being ridden like a horse by the inamorata of Alexander the Great.

Chapter 3 focuses on the vastly increased use of animals in clerical attempts, particularly those following on from the Fourth Lateran Council, to improve the theological education and moral behaviour of the laity. Stories about animals and their characteristics proved extremely useful to this enterprise, sweetening didactic pills, facilitating the comprehension of unfamiliar or difficult ideas and bringing home moral messages in down-to-earth ways that all could relate to. I argue, however, that, far from supplanting older medieval traditions of using animals to convey religious and/or ethical meanings, the "new" Aristotelian ways of looking at the natural world renewed and enriched them. This can be seen for example in the Dominican encyclopaedist Thomas of Cantimpré's habit of providing his readers both with fresh information about exotic species and with allegorical interpretations of their behaviour which could be slotted neatly into sermons; in the development of systematically ordered collections of nature exempla; and in the remodelling even of some of the most ancient animal allegories to take account of newly observed ways in which the creatures in question actually behaved.

Chapter 4 turns away from explicitly learned environments to explore various functions and portrayals of animals in the context of thirteenth-century chivalry. It focuses particularly on the many and diverse attempts made by members of the lay aristocracy to define or redefine their identity through the medium of animals – whether as symbols of status or wealth, heraldic emblems, or companions in peace and war. Particularly through a discussion of knights' relationships with their horses as these are portrayed in vernacular courtly romances, the chapter also reveals the remarkable extent to which some thirteenth-century lay aristocrats were ready to question and challenge traditionally held conceptions of the "anthropological difference" between humans and animals. Sometimes, as I show, human and equine identities are presented as so intertwined and mutually dependent as to seem, in effect, to merge.

Finally Chapter 5, which is built on sources from a wide variety of social contexts, foregrounds a tension that was as characteristic of the thirteenth-century Animal Turn as it is of its twenty-first-century equivalent – namely that between violence towards animals and affection for them (the dilemma of the carnivorous pet-lover, as it were). The chapter presents St Francis of Assisi as having been prey to a perhaps surprising degree to this tension, and goes on to explore several examples of explicit violence, as well as evidence that pets in particular were treated with increasing levels of affection. Overall it seems highly likely that acts of violence towards animals remained much more common in the thirteenth century than expressions of affection for them. Even here, however, one can observe some signs of change: in contexts where violence towards animals was at least implicitly equated with violence towards humans, or where the lot of animals was materially improved by technological change. A key example of this is the discovery of new papermaking techniques in thirteenth-century Italy, which led to a gradual decline in the use of parchment, and hence in the mass murdering of calves to make books.

Throughout this discussion a point is made of consulting a wide variety of textual sources: these include epics, romances, lyric poems, natural historical studies, encyclopaedias, sermons, collections of nature exempla, travel literature, saints' lives and a treatise on falconry. It is anticipated that some of these will be familiar to many readers, and others less so. The latter is likely to apply particularly to several works (discussed mainly in the later chapters) that were written in the German-speaking lands. If this means that it is a side effect of my argument to enhance the profile of primary sources that deserve to be better and more widely known, that is of course all to the good.

IV

So where does this volume fit within, and how might it be said to extend, the parameters of existing research into medieval animals? Self-evidently, first of all, it takes account of – and responds to – the many important contributions made in recent years by medievalists working within the area known as Animal Studies or (not least in Germany) Human-Animal Studies – "an interdisciplinary field that explores the spaces that animals occupy in human social and cultural worlds and the interactions humans have with them. Central to the field is an exploration of the ways in which animal lives intersect with human societies" (DeMello 4).

The earliest of these contributions was that of Joyce E. Salisbury, the first edition of whose book *The Beast Within* appeared in 1994 and has proved highly influential. A particular and, for its time, unusual strength of Salisbury's volume is the fact that she discusses both real-life and fictional animals. She begins by focusing on animals as property, and hence as sources of labour, materials, food and in some cases illicit sex; but she goes on to analyse the use of animals in literature – especially fables and bestiaries – as exemplars and metaphors for human behaviour. Her book's underlying theme is expressed at the very beginning: "by the late Middle Ages (after the twelfth century), the paradigm of separation of species was breaking down" (1). The evidence given for this is perhaps stronger in some chapters than others, and the precise contours of Salisbury's chronology can be vague: the phrase just quoted as "after the twelfth century" elsewhere becomes "in the twelfth century"; Gerald of Wales is variously described as a twelfth- and a thirteenth-century author (1, 75); the phenomenon of ladies keeping lapdogs is said to have originated in the twelfth century and also in the thirteenth (ix, 118); and the first case of an animal trial, presented as a feature of the twelfth and thirteenth centuries (108), is dated by Salisbury (correctly) to the 1260s. I am going into such pernickety detail because one of the ways in which the present book seeks to build on Salisbury's work is to be a little more precise about when attitudes to animals in medieval Europe began to turn. My argument is that this happened in the thirteenth century; but some of Salisbury's arguments make it helpfully clear that a propitious climate for innovative thinking had been created already in the decades prior to 1200.

Her sense of the sometimes porous distinctions between the human and the animal has proved a particularly fruitful stimulus for other medievalists. Among the first of these was Dorothy Yamamoto (2000), who focuses on various kinds of body and the ways in which they were valued in the Middle Ages: human and animal bodies, high and low ones, central and peripheral (or marginalized) ones. She shows that, contrary to expectations, humans are not always thought of as "high" and "central" and animals as "low" and "peripheral". Birds, for example, frequently embody various forms of noble behaviour; hunting dogs inhabit a taboo space where they are neither man nor entirely animal; and, not least in *Sir Gawain and the Green Knight*, human and animal bodies are united in their essential shared mortality. Meanwhile in German, Udo Friedrich (2009) examines numerous theological, political and literary discourses surrounding the often fluid medieval frontiers between humanity and animality. He demonstrates not least that such discourses can clash creatively in thirteenth-century literature: for example in *Wolfdietrich A* (c. 1230), in which the protagonist must overcome elements of animality both within himself and in his opponents before arriving at a decidedly Christian form of heroic maturity, and in Konrad von Würzburg's *Partonopier und Meliur* (1270s), in which the idealized presentation of the hero is both undermined and enriched by a series of animal similes and metaphors.

In more recent work operating within the framework of Animal Studies, the problematic human/animal divide tends to function more as a kind of intellectual undergirding, rather than constituting the prime focus of thematic attention. Karl Steel (2011), for example, foregrounds the issue of violence perpetrated by humans on animals (see Chap. 5 below). Susan Crane (2012) discusses various encounters, real and imaginary, between humans and animals: in the cohabitation of saints and animals; in the taxonomy and interpretative arsenal of bestiaries; in the ritualized cut and thrust of the hunt; and, not least, in the mechanical and affective aspects of the relationship between knight and horse (see also Chap. 4). Sarah Kay (2017) examines parchment bestiary manuscripts to establish links, or "sutures" between their animal subject matter, their human readers and the animal skins on which the texts are written and illustrated (see Chap. 5). Finally Peggy McCracken (2017) looks at literary human-animal encounters with a view to exploring medieval conceptions of human sovereignty and authority – whether it be wielded over animals or fellow humans. In this context she covers representations of flayed animals, the sometimes consensual nature of animal domestication and human subjugation, issues of self-sovereignty and the extent to which sovereigns and animals can share an identity or a symbolic kinship.

As will be seen, the present study makes extensive use of this recent research, as well as contributions by other scholars who do not explicitly present their work as "Animal Studies", but nevertheless share several of its approaches and priorities – I am thinking here, for example, of David Salter (2001), Peter Dinzelbacher (2000, 2002) and Jeffrey J. Cohen (2003).

Taken as a whole, there are several things that medievalists' contributions to Animal Studies have done particularly well. They have kept a clear but flexible focus on the many interfaces at which humans and animals meet; they have drawn many fruitful links between "actual" animals and animals as portrayed in literature and art; they have made appropriate and productive use of literary and cultural theory; they have consistently worked in ways that transcend conventional disciplinary boundaries; and, above all, they have helped scholars of many specializations to form the habit of looking *at* animals, rather than only *through* them, in search of some putatively more important abstract meaning.

There are, however, also certain things at which medieval Animal Studies is rather less good. Maybe inevitably, given the centrality to the field of actual contact with humans, it has tended to work with a relatively limited repertoire of animals which medieval Europeans either encountered in real life or often read about – horses, dogs, wolves and to a lesser extent falcons, bears or lions. Similarly, medieval Animal Studies is apt to take account only of a rather limited range of source texts, especially bestiaries, beast epics, the fables and *Lais* of Marie de France, Chaucer and French or English romances. As a result much important medieval writing in Latin and, particularly, in German has tended to play at best a peripheral role.

In this context the present volume seeks to make a virtue of concentrating on a rather different and somewhat wider corpus of primary texts. Self-evidently many choices of what works to write about have been made for me by the decision to focus, as rigorously as possible, on the thirteenth century. If I had been writing mainly about the twelfth century, for example, both beast epic and the French courtly romance would have been more important for my argument than they are; and if my focus were on the fourteenth, I would have had to say much more about Dante and Chaucer. Nevertheless it is no coincidence that a book about the thirteenth century should use a significant number of sources that originated in what we would now call Germany. In that century (especially, though by no means only, if one were to cheat by making it begin around 1180)

German literature underwent a remarkable flowering – with regard both to the great works of lyric poetry and narrative fiction that can be assigned to its early decades, and to the diversity and quality of didactic literature, both in Latin and in the vernacular, that adorned much of its course. It is therefore important that the contribution of Germany (in the term's widest sense) should receive an appropriate – though one hopes not disproportionate – level of coverage in a work of this kind.

With regard to its methodology, this volume makes a concerted attempt to combine and reconcile approaches that can sometimes seem to be at variance. That is to say, I try to use what I have learnt from the work of Animal Studies scholars, whilst at the same time bringing to bear the knowledge and methodological emphases I myself have acquired over the course of a somewhat different academic formation – as a medieval Germanist with a strong interest also in Latin literature, particularly of a moral theological and/or encyclopaedic orientation. My work specifically on animals has owed a great deal to an ongoing German tradition of studying the allegorical use of animals to convey spiritual meanings. This was inaugurated on a theoretical basis by Friedrich Ohly, in a seminal article from 1958 which finally appeared in English in 2005, and on a practical one by Dietrich Schmidtke, whose monumental 1966 thesis on the spiritual interpretation of animals in German literature between 1100 and 1500 has never been published, let alone translated. Nevertheless these two initiatives have been influential in stimulating many kinds of subsequent work – catalogues of animal meanings (Henkel, Lecouteux, Harris 1994); major monographs on individual animals (Einhorn, Gerhardt, Schumacher); substantial editorial initiatives (Helmut Boese, Benedikt Konrad Vollmann, Georg Steer and myself); and extensive work on Latin and German nature encyclopaedias (Christel Meier, Heinz Meyer, Gerold Hayer, Baudouin van den Abeele).

Whilst all of this scholarship deserves to be much better known amongst English-speaking readers, it too, of course, has its deficiencies. For one thing, with the partial exception of the encyclopaedias, all the texts it considers could be said to fall victim to what Roland Borgards (2015, 156) calls the "snare of anthropocentrism" – that is to say, they think of animals primarily as signifiers that refer to aspects of the human or divine world, rather than as actors, individuals or agents in their own right. Nevertheless, such a way of looking at creation undeniably remained widespread throughout and beyond the Middle Ages, and needs to be taken into account if our understanding of the culture(s) of the period is to remain accurate. Moreover real animals and literary or symbolic ones do not and cannot exist in mutually exclusive vacua. There may be a distinction between them, but, as Borgards again says (2015, 156), "this distinction is by no means self-evident, trivial, natural or easy. Real wolves seem to play a formative role in every literary wolf – a process requiring serious scholarly attention – and every real wolf likewise exhibits traces – equally meriting careful scholarly attention – of their literary counterparts". In Virginie Greene's term, then, symbolic animals, whether in a book or on a shield, are often adstractions of real animals – a theme to which we will return in later chapters.

V

Before embarking on the main thrust of the book's argument, three clarifications (or indeed admissions) are needed – one involving time, one geography and one terminology.

Firstly it needs to be stated that, in spite of this book's title, I can offer no assurance that every source referred to in it as "thirteenth-century" necessarily originated between the years 1200 and 1300. It is of course ultimately invidious to try to press a long-term and organically developing cultural phenomenon such as an Animal Turn into an artificial straitjacket bounded by two calendar years; and with regard specifically to the Middle Ages, it is also impossible to do this with any guarantee of accuracy: so many dates are either approximate, or contested, or both. In the light of this it is tempting to take refuge in the notion of a "long thirteenth century", by analogy with similar constructs used by historians of the eighteenth and nineteenth centuries. On the whole, however, I have avoided using such a concept: partly because it too would necessarily be compromised by the uncertainties inherent in all dating processes relevant to the Middle Ages; and partly because there could never be any agreement as to when a "long" thirteenth century might begin and end. In the case of, say, the "long nineteenth century", there are obvious and persuasive reasons why most people would regard it as beginning in 1789 and ending in 1914; but around the years 1200 and 1300 there were no such cataclysmically influential and securely datable events as the French Revolution and the start of the First World War, and hence there is no realistic prospect of a comparable scholarly consensus emerging. On balance it is probably better to use the straightforward term "thirteenth century", on the understanding that it is not a watertight or invariably apt category. Unless otherwise stated, then, all the texts described in the chapters that follow as dating from the thirteenth century (and that is most of them) are generally held to have been composed between 1200 and 1300, give or take only a very few years on either side; but absolute certainty about this is in many cases impossible.

Secondly, it must be owned that the scope of this volume, whilst encompassing works produced in several medieval languages, is consciously Eurocentric. Indeed, most of the works I have read in preparing it originated in France, the Holy Roman Empire, England or Italy. There are exceptions to this: Aristotle was self-evidently a Greek author, and Michael Scot made his all-important translations of him in Toledo; and we also hear accounts, from writers such as Marco Polo and Jacques de Vitry, of life beyond the borders of Europe. On the whole, though, we do not venture very far beyond what one might call the medieval West; and I am conscious that, if we had done, different perceptions might have been registered and different conclusions drawn. Maybe they will indeed be brought to bear in due course – but for that we shall need a different book, by a different and suitably qualified author.

Finally to an aspect of terminology. Throughout this volume, human animals are designated "humans" and non-human animals "animals". This, of course, is intrinsically unsatisfactory, particularly in the context of a discussion which will frequently point to the essential fluidity and questionability of the two terms. They are not used blithely or unthinkingly, though, but only because of the lack of a viable alternative. It is comforting to note that Susan Crane has experienced something of the same frustration about this matter: "It is hardly helpful to resort to scare quotes around "animal" or to new locutions such as the *arrivant,* the *strange stranger,* the *animetaphor* and the *animot:* these locutions have made important points about one or other problem with the *animal,* but none can confront all its inadequacies" (2). Quite. So "human" and "animal" it is – with, of course, all the necessary reservations and caveats.

CHAPTER 2

Aristotle and Thirteenth-Century Animal Studies

Abstract This chapter argues that the zoological works of Aristotle, newly translated into Latin by Michael Scot, were a seminal factor in the thirteenth-century Animal Turn. They provided Western intellectuals with new scientific information, but above all with new ways of looking at animals – as objects of serious study in their own right, and as phenomena to be learnt about by observation and rational judgement, rather than by consulting the works of ancient authorities. Several thirteenth-century works inspired by Aristotle are examined, especially the *De animalibus* of Albertus Magnus and the *De arte venandi cum avibus* of Emperor Frederick II. This analysis considers not just the zoological content of such works, but also their structure and contributions to ongoing debates about the human-animal divide.

Keywords Albertus Magnus • Aristotle • Medieval science • Zoology

I

The translation into Latin of the surviving corpus of works by Aristotle (d. 322 BC) was a seminal development in the history of European thought. This is true not least of his writings on natural history, the pattern of whose medieval reception was a diverse and at times eccentric one – but one which had the cumulative effect of radically changing the ways in

© The Author(s) 2020
N. Harris, *The Thirteenth-Century Animal Turn*,
https://doi.org/10.1007/978-3-030-50661-2_2

which animals and other natural phenomena were observed, perceived, interpreted and used to construct meaning.

Of the two principal high-medieval Latin translations of Aristotle, by far the more influential was that prepared in Toledo, almost certainly in the second decade of the thirteenth century, by Michael Scot. His *De animalibus* translates Arabic versions of Aristotle's *Historia animalium, De partibus animalium* and *De generatione animalium*. The extent of this translation's success in the Middle Ages is attested in part by its transmission in more than sixty surviving manuscripts (considerably more than contain the later thirteenth-century version by William of Moerbeke), but above all by the number of authors who demonstrably used it. These include commentators concerned to explain Aristotle to a new audience; authors of glosses focusing on specific words or problems; compilers of paraphrases, *abbreviationes,* compendia or florilegia seeking to present Aristotelian material in an accessible, manageable form; and writers of "independent treatises or monographs dealing with special problems in Aristotelian philosophy" (Lohr 313).

Very many of these users of Aristotle were active in the years between 1220 and 1300. These decades saw the production, for example, of the Aristotelian florilegium *Parvi flores* of Johannes de Fonte, the *Questiones super XVIII libros de animalibus* of Roger Bacon, the *Liber de animalibus* of Pedro Gallego and, perhaps most importantly, the great nature encyclopaedias of Thomas of Cantimpré, Bartholomaeus Anglicus and Vincent of Beauvais.

Amongst this plethora of thirteenth-century voices influenced by Aristotle, however, two can be singled out as being of particular, paradigmatic significance: those of the Cologne-based philosopher, theologian and natural scientist known as Albert the Great (Albertus Magnus), and of the Holy Roman Emperor Frederick II. Albert was almost certainly the single most important reader, interpreter and teacher of Aristotle in the thirteenth century; and in the case of the zoological treatises in particular, his reception of them exceeded all others' "in both quality and quantity" (van den Abeele 1999a, 303). Above all, Albert's massive 26-book *De animalibus,* completed around 1260, is the longest and most original medieval commentary on and development of Aristotle's natural historical thought. For his part, Frederick II facilitated the spread of Aristotelian ways of thinking by commissioning Michael Scot to translate Avicenna's *Liber de animalibus* – based closely on Aristotle – but above all by composing a wholly remarkable volume on the subject of falconry (and also, in practice, much more), the *De arte venandi cum avibus* (1240s). The first of its six books in particular bristles with obvious indebtedness to Aristotelian methodology, and the compendious treatment it accords to numerous bird species can be seen as making good Aristotle's tendency to avoid avian subjects at the expense of species who dwell on the ground or – especially – in the sea (van den Abeele 1999a, 311). If the discussion which follows focuses especially – though not exclusively – on these two central thirteenth-century works, it is because in their different ways they constitute uniquely creative and innovative assimilations of the "new", Aristotelian science of animals, and are as such particularly eloquent witnesses to the thirteenth-century Animal Turn.

II

In what precise ways, then, did Aristotle promote such a turn? His influence is much more often asserted than it is analysed; but my contention is that it can be seen above all in five ways: in the provision of information about animals that had remained unheeded for centuries; in a pervasive, sometimes all-consuming thirst after knowledge for its own sake; in a reliance on observation and reason, as much as on the writings of earlier authors; in an urge to systematize natural historical and zoological knowledge; and, not least, in a distinctive – if somewhat ambiguous – view of the so-called anthropological difference separating animals from humans.

It must be stated at once that knowledge of Aristotle's writings on animals had never died out completely, but rather was transmitted throughout antiquity and the Middle Ages – in the form of quotations, paraphrases and summaries – via such intermediaries as Pliny, Solinus or Isidore of Seville. By definition, however, such transmission could only be indirect and partial, and there is no doubt that the direct access to his work furnished by the thirteenth-century Latin translations meant that much more of the scientific information Aristotle provided became widely known in the West. The extent and swiftness of the infiltration of Aristotelian material into the great thirteenth-century Latin nature encyclopaedias can easily be traced. Whilst in the *De naturis rerum* (c. 1190) by the Augustinian Alexander Neckam Aristotle is little more than an occasionally cited source amongst many, by 1240 we find him figuring far more prominently in the *De proprietatibus rerum* of Neckam's Franciscan compatriot Bartholomaeus Anglicus and, especially, the *Liber de natura rerum* by the Dominican pupil of Albertus Magnus, Thomas of Cantimpré. In his prologue, the latter places Aristotle first in the list of *auctoritates* he has used, and describes him as having "blossomed forth more eminently than all others not only in these matters, but in all things that pertain to the discipline of philosophy" (ed. Boese, 3). In what follows, Aristotle transpires to be Thomas's third most frequently cited source, after the previously ubiquitous Pliny and the shadowy, still unidentified *Experimentator* (van den Abeele 1999a, 308).

Numerous details make it clear that Thomas, and hence Albert (whose Books XXII–XXVI are based on his pupil's encyclopaedia) drew directly from Aristotle. It is only to him, for example, that they can owe their knowledge of various fish species, such as the *alforaz*, which is capable of both generation and regeneration upon contact with water or mud (Albert, trans. Kitchell/Resnick, 1661); or the *celethi*, which, "because of the weight of its head, sleeps so deeply that it can be caught by hand when it is asleep" (1677); or indeed the extraordinary *hahane* (1664), whose "stomach swells out beyond all reason" and, when it fears danger, "folds its skin and fat over its head, hiding its head like a hedgehog" – only to start eating this skin and fat if the danger remains long enough for it to get hungry. Paradoxically, the very strangeness of such material confirms that it did not come through the standard Latin intermediaries: the species names are corruptions of Greek or Arabic terms (a common problem amongst medieval readers of Aristotle), and some of the more bizarre details stem from demonstrable misunderstandings of the *Historia animalium*.

On one level, then, Aristotle was seen as a significant source of factual information about the natural world; and indeed, as one worthy of more trust and respect than some others. It is noteworthy that Albertus Magnus, whilst by no means always agreeing with what he has read in Aristotle, nevertheless takes pains to disagree politely, indirectly and/or implicitly. "Aristotle and certain other philosophers", he cautiously claims, "place the eagle and the vulture in the same genus of bird... On account of this, some say that the vulture is the most noble genus of bird. This is not our usage, however, for in our lands the vulture is a very big, lazy, ignoble bird" (1551). When dealing with Roman authorities such as Pliny, by contrast, his quill is very much sharper. Pliny has claimed that there exists a one-eyed species of heron, which he rather unimaginatively calls the *monoculus*. "But", Albert objects, what he says is false and contrary to nature. For, just as two wings and two feet grow from the sides, so do two eyes. Reason does not allow that one eye is formed from one side and another is not formed from the other side. To be sure, this Pliny says many things that are entirely false and in such matters his words should not be given consideration (1556).

He never speaks like that about Aristotle.

III

Of considerably greater importance than his provision of often recondite lore, however, was the infectious influence Aristotle had on thirteenth-century attitudes towards the natural world. First among these was an insatiable thirst for knowledge in its own right and for its own sake. Jonathan Barnes has declared that "throughout his life Aristotle was driven by one overmastering desire – the desire for knowledge. His whole career and his every known activity testify to the fact: he was concerned before all else to promote the discovery of truth and to increase the sum of human knowledge" (3). An unbounded desire to gain and to pass on knowledge is reflected also in the sheer range and scale of Aristotle's intellectual enterprises. Zoology was, after all, only one of his many interests, which encompassed also philosophy, politics, mathematics, biology, the arts, history, law – indeed, "choose a field of research, and Aristotle laboured in it; pick an area of human endeavour, and Aristotle discoursed upon it" (Barnes 4).

There is no doubt that such ambitious enthusiasm rubbed off on, and chimed in with, the preoccupations of many thirteenth-century intellectuals. Theirs was, after all, the age of the encyclopaedia, with its aspirations to completeness, to universality – aspirations which were not confined to the written word, but also to such other forms of activity as map-making. It is no coincidence that the Ebstorf and conceivably also the Hereford *mappae mundi* date to the thirteenth century.

Underlying such enterprises, just as it had underlain the indefatigable scientific scholarship of Aristotle, was a drive to know not only about things in general, but about the natural world in particular. In many respects this was new. Medieval people had always been interested in animals, but in the main as sources of food, transport or companionship, rather than as subjects for what we would now call scientific study. And intellectuals, following consciously or unconsciously in the footsteps of St Augustine, had tended to view the study of natural phenomena as a means to a theological or moral end, rather than a worthwhile scholarly enterprise in itself. In the familiar phrase, nature was in essence a book, whose prime functions were to reveal and reflect the glory of God, and to show us as human beings how to behave. Thomas of Chobham (c. 1160–c. 1235) sums up this remarkably anthropocentric view of nature as well as anyone:

> The Lord created different creatures with different natures not only for the sustenance of men, but also for their instruction, so that through the same creature we may contemplate not only what may be useful to use in the body, but also what may be useful in the soul... there is no creature in which we may not contemplate some property belonging to it which may lead us to imitate God, or some property which may move us to flee from the Devil. For the whole world is full of diverse creatures, like a manuscript full of different letters and sentences in which we can read whatever we ought to imitate or flee from (quoted from Camille, 355).

Even some thirteenth-century encyclopaedists reveal an indebtedness to this way of seeing the world. In the prologue to his *De proprietatibus rerum,* for example, Bartholomaeus Anglicus states that his purpose in writing is to help people "understand the enigmas of scripture which have been passed on and hidden by the Holy Spirit under symbols and figures from the natural world", in the knowledge that "the mind cannot ascend to the contemplation of invisible things unless it has been led to contemplate visible things" (6^r). Whether such statements reflect genuine conviction or merely a desire to do lip-service to tradition is hard to say – the latter is undoubtedly possible, given that Bartholomaeus was writing at a time when Aristotle was a controversial figure in some circles. What is certain, however, is that his nature encyclopaedia actually goes far beyond such a spiritual agenda. It is as though Bartholomaeus's interest in nature per se simply takes over as he writes – as though, one might go so far as to say, an Animal Turn begins spontaneously to take shape under his pen. This is observable in the other thirteenth-century encyclopaedists also: there is a new sense of curiosity, of a passion to know things, that really does seem to have been set in train by prolonged exposure to Aristotle.

That this burgeoning attitude was not solely the preserve of churchmen is moreover revealed by the *De arte venandi cum avibus* of Frederick II. The minutely detailed character of his work, as well as its sheer size, bespeak an avid interest in birds qua birds that far transcends such conventional aristocratic preoccupations as their usefulness for fine dining or courtly display. In his prologue Frederick tells us something about his level of commitment: "We have investigated and studied with the greatest solicitude and in minute detail all that relates to this art, exercising both mind and body so that we might eventually be qualified to describe and interpret the fruits of knowledge acquired from our own experiences or gleaned from others" (Frederick, trans. Wood/Fyfe 3). Nothing in the extensive treatise which follows leads us to doubt the veracity of these words.

IV

Nature, then, in spite or maybe because of its unmanageable complexity, was becoming – perhaps for the first time since Greek antiquity – a serious subject for study in its own right. But how precisely was knowledge about it to be acquired and added to? Not, in the main, by using the full range of methods available to modern experimental science. Extravagant claims have been made with regard to the modernity and originality of both Aristotle and Albertus Magnus. It is something of a cliché to refer to the former as the "father of modern science"; and the latter also has not infrequently attracted glowing tributes, such as the one found in the current edition of the *Encyclopaedia Britannica,* to the effect that he merits "a preeminent *[sic]* place in the history of science". There is a good deal to be said for such claims; but we must be careful not to overstate the case or to make it lazily.

Neither Aristotle nor Albert was, or wished to be, an experimental scientist in the modern sense. They observed nature in the field, rather than testing hypotheses in a laboratory. Undoubtedly they did both undertake a certain number of experiments. In Aristotle's case, these seem to have consisted mainly of dissections and vivisections of animals in order to gain information about their bodily functions and structures. Sadly, an illustrated volume by him entitled *The Dissections* (*Anatomata*) has not survived. Albert's experiments may have been less extensive, interventionist or indeed morally questionable, but he certainly did perform some. In *De animalibus,* for example, he tells us that, in order to interrogate the time-honoured belief that ostriches eat and digest iron, "I have often spread out iron for several ostriches and they have not wanted to eat it. They did greedily eat rocks and large, dry bones that were broken into smaller pieces" (1648). One trusts that no ostriches suffered lasting harm.

Something rather closer to modern scientific experimentation, with its insistence on controls and repeatability, can occasionally be found in Frederick's *De arte venandi*. The Emperor, we are told, wished to investigate the notion that certain birds, again including the ostrich, neglect to incubate their own eggs, relying instead on the heat of the surrounding climate to hatch them. "A similar phenomenon", he avers, "is to be observed in Egypt, where eggs of the barnyard fowl are kept warm and the young hatched out independent of the mother bird. We ourselves saw this, and we arranged to have it repeated in Apulia by experts whom we summoned from Egypt" (53). Such a commitment to concerted enquiry and to the proactive pursuit of corroborative evidence was however rare in the thirteenth century (and indeed for a considerable time beyond). Few indeed were those who possessed the requisite time, resources and indeed intellectual curiosity brought to bear by Frederick on his investigation of the legend of the barnacle goose:

> There is… a curious popular tradition that they spring from dead trees. It is said that in the far north old ships are to be found in whose rotting hulls a worm is born that develops into the barnacle goose. This goose hangs from the dead wood by its beak until it is old and strong enough to fly. We have made prolonged research into the origin and truth of this legend and even sent special envoys to the North with orders to bring back specimens of those mythical timbers for our inspection. When we examined them we did observe shell-like formations clinging to the rotten wood, but these bore no resemblance to any avian body. We therefore doubt the truth of this legend in the absence of corroborating evidence (51f.).

For all their dedication, skill and historical significance, then, we really should not think of even the greatest scientists of the ancient and medieval worlds as pioneering experimental researchers in a modern sense. That does not mean, however, that their work was anything other than remarkable and innovative. Certainly the natural history produced by Aristotle, Albert and Frederick was quite radically different in character from that standardly practised in the Roman period or the earlier Middle Ages. It is not anachronistically simplistic to claim that Aristotle's programme of biological and zoological enquiry had been – for all his continuing presence in collections of quotations – largely ignored during these periods. Writing about animals was generally left to theologians or other non-experts, who were content in the main simply to collate and repeat uncritically the zoological or pseudo-zoological findings of earlier ages. Quite remarkably little progress was made between, say, the third century BC and the thirteenth century AD.

Nevertheless a genuine, sometimes exaggerated respect for the writings of venerable authorities was not exclusively confined to what one might think of as the zoological "dark ages". Aristotle himself was in no way dismissive or disrespectful of the work of those who had gone before him. Indeed, he was fully conscious of the debt he owed them. Much the same could be said of any of the thirteenth-century writers on nature we have been alluding to or quoting from. For all of Albertus Magnus's occasional barbed comments about Pliny or Solinus, for example, he continues to use them extensively as sources, to the extent of transmitting – to us – plainly mythical lore about creatures such as the manticore, the phoenix or the basilisk without discernibly batting an eyelid. Frederick II, too, is notably respectful towards the writings of his principal source, in this case Aristotle. He makes frequent references to Aristotle being correct, leaving nothing essential unsaid, and the like; moreover he even writes in his first chapter, "all that we do not include on the nature of birds can be found in Aristotle's book *On animals*" (7).

With Frederick, though, things are not always as straightforward as that. He often seems ahead of his time in asserting a more thoroughgoing degree of independence than that claimed by any other medieval writer on the natural world I have read. Even Aristotle can err; and when he does, Frederick can be relied upon to say so. Aristotle, he says, was wrong to call raptorial birds "greedy-clawed" or "birds of the hooked claws", since "birds such as jackdaws, the larger swallows, and vultures have hooked claws and yet may not properly be called raptores" (9); and, in his prologue, Frederick – as ever employing the royal "we" – programmatically declares that:

> We discovered by hard-won experience that the deductions of Aristotle, whom we followed when they appealed to our reason, were not entirely to be relied upon, more particularly in his descriptions of the characters of certain birds. There is another reason why we do not follow implicitly the Prince of Philosophers: he was ignorant of the practice of falconry... In his work, the *Liber animalium*, we find many quotations from other authors whose statements he did not verify and who, in their turn, were not speaking from experience. Entire conviction of the truth never follows mere hearsay (3f.).

This passage can be seen as both Aristotelian and forward-looking. The work of those who have gone before is of value and not to be dismissed lightly; but it must be tested and verified by means of observation and experience, and assessed on the basis of reason.

In spite of Frederick's criticism of him here, this is in essence the way in which Aristotle himself worked. He viewed the pursuit of knowledge as a collective labour in which tradition played a literally foundational role. At the same time, he maintained a certain critical distance to such handed-down material, regarding it not as definitive, but rather as a necessary starting point for his own enquiries. And these enquiries were nothing if not empirical, founded as they were, if not strictly speaking on repeatable experimentation, then on an indubitably potent combination of observation and reason.

Aristotle was an inveterate observer of the natural world. Legend has it that he was encouraged in these pursuits not least by the insatiable curiosity of his pupil Alexander the Great. Pliny, no less (VIII, xvi, 44), claims that "inflamed by a desire to know the natures of animals", Alexander appointed "several thousand men throughout the whole of Greece and Asia Minor to be at Aristotle's disposal – everyone who lived by hunting or falconry or fishing, or who looked after parks, herds, apiaries, fishponds, or aviaries – so that no living creature should escape his notice" (quoted from Barnes, 24). This is hard to take at face value, but there is no doubt that Aristotle did seek to look into as much of the natural world as was humanly possible, and that he did rely on the testimony of expert witnesses "on the ground". As with his written sources, these informants were treated with respect but in no sense with gullibility. So when someone claims that fish do not copulate, Aristotle is swift to disagree, charitably but firmly: "Their error is made easier by the fact that such fish copulate quickly, so that even many fishermen fail to observe it – for none of them observes this sort of thing for the sake of knowledge" (Barnes 24f.).

A strikingly similar approach is adopted in the thirteenth century by Albertus Magnus. He too is constantly aware of the work of earlier authors. Indeed, he often makes a determinedly systematic effort to sift, compare and explain the opinions of his scholarly predecessors in some detail before arriving at his own conclusion. This means that, for example, his consideration of the rise of the veins in animals becomes a highly complex exercise – which begins with Aristotle, but ends with Albert's own, by now thoroughly considered and tested opinion:

Since it is our intention to pass on perfect doctrine in these matters, we will first set out the opinion of Aristotle on this matter, with a clear explanation. Second, we will point out what Galen says to the contrary, and third, we will introduce the solution of Avicenna on these matters. Fourth, however, we will point out all that Averroes says that is contrary to Avicenna and we will set forth his solution. Fifth, and finally, we will educe our opinion from all these, and we will prove it by use of reason and solid experiential knowledge that is completely trustworthy (351).

In addition to written sources, Albert also was plainly keen to use the evidence of his own eyes. One can readily imagine him, as one can Aristotle, assiduously observing and noting down various natural phenomena he witnessed, either at home or on his extensive journeys in the service of the Dominican Order. It was personal observation, for example, that convinced him that humans could safely eat spiders:

> Moreover, in Cologne, and in our presence and in front of many of our brethren, it happened one time that a girl of three, as soon as she was set down from her mother's arms, ran about along the corners of the wall seeking spiders and ate them all, large and small. Further, she did well on this food and desired it over all others (648).

Furthermore, like Aristotle before him, Albert seems to have used an extensive network of experienced witnesses ready to furnish or confirm natural historical information. He frequently quotes from these, but, like Frederick II, seems particularly happy when he himself has been present to corroborate and assess their evidence. This is why he can be confident of his knowledge of the mating habits of fish:

> While diligently observing and making inquiries of the oldest fishermen of the sea and rivers, I have seen with my own eyes and heard with my own ears that during mating time fish rub their bellies together and spread their eggs and milt alike at contact (496).

On occasion one is inclined to wonder whether Albert's commitment to empirically ascertained truth is quite as uncomplicated as it initially seems. John B. Friedman (1997), for example, has argued that Albert's tendency to "nationalise" or "regionalise" natural historical data by inserting brief references to specifically German natural phenomena should be regarded with an element of suspicion. This is in essence a rhetorical device, Friedman suggests, designed to "create an experiential tone" which might mislead Albert's readers into thinking that his own in the flesh experience of the animals he discusses was more extensive than it really was. There is something in this. Friedman points out, for example, that Albert's treatment of the *hemptra* (Alpine marmot? hamster?) is taken entirely from Thomas of Cantimpré save for Albert's assurances that the animal is "found in Germany" and that it is "found only in mountainous regions and is the largest member of the rodent family in our country" (Friedman 389). To this one could add numerous other brief statements to the effect that a certain thing obtains or has occurred "in Germany", "in our lands", or indeed "in Cologne". These hardly inspire confidence in the reader that they are based on any extensive research on Albert's part. One should not overstate the case, though. Even Albert's toponym-dropping often has the ring of truth about it, especially when he provides a little more by way of geographical detail. Pliny is wrong, Albert asserts, to state that vultures do not nest; and this can be shown to be false because "vultures build nests every year in the mountains which are between Trier and the Civitas Wangionum, which is now called Worms. So much do they do this that the land reeks from the carcasses carried there" (1654). This reads much more like a genuine attempt to add to knowledge than an attempt to manipulate reader responses by means of rhetorical skill. So does Albert's account of the attempts made by "the country folk of Bohemia and Carinthia" to capture food-gathering pre-hibernation dormice by means of "little storehouses in the forest". In the Autumn, it seems, dormice "take up residence in these in very great numbers and then are collected in them for human consumption" (1510).

What in the end makes it possible and plausible to take most of what Albert tells us at face value is the fact that his – Aristotelian – blend of critical respect for authorities, reliance on observation and use of reason as his final arbiter was by no means unique to him as the thirteenth century progressed. New questions, perspectives, habits were beginning to develop – sometimes in perhaps unexpected places. A good example of this from the very end of the century is the celebrated Venetian traveller Marco Polo. His *Travels* certainly reveal him to have been a skilled raconteur and rhetorician, but he was plainly also an acute and inquisitive observer of the natural world. Many times Polo conveys information which could have been gleaned only from direct contact with animals and a keen interest in them. He was clearly fascinated, for example, by – to a Westerner – unfamiliar ways in which animals were actually used in the thirteenth-century Orient. There are many references to camels as beasts of burden, or to elephants as participants in war. He also observes and records less quotidian uses of animals to facilitate travel: wolves apparently kill many wild sheep and goats in the province of Vokhan, such that, the latter species' horns and bones being found in large quantities, "heaps are made of them at the side of the road, for the purpose of guiding travellers at the season when it is covered with snow" (49). Perhaps most importantly for our current discussion, however, Marco Polo uses critical observation to extend and even revise the zoological knowledge recorded in ancient and medieval encyclopaedias. *Histrix,* the porcupine, is for example alluded to briefly by naturalists from Pliny to Albert the Great, but its unique defence mechanisms were not recorded in any detail before Marco. "Here are found porcupines" – he says of the town of Scassem (Ishkāshim? 43) – "which roll themselves up when the hunters set their dogs at them, and with great fury shoot out the quills or spines with which their skins are furnished, wounding both men and dogs" (43). Most significantly of all perhaps, Polo's encounter with a rhinoceros in the Kingdom of Basman enables him to clear up, potentially definitively, the ancient scholarly confusion between this animal and the unicorn:

> In the country are many wild elephants and rhinoceroses, which latter are much inferior in size to the elephant, but their feet are similar. Their hide resembles that of the buffalo. In the middle of the forehead they have a single horn; but with this weapon they do not injure those whom they attack, employing only for this purpose their tongue, which is armed with long, sharp spines, and their knees or feet; their mode of assault being to trample upon the person, and then to lacerate him with the tongue. Their head is like that of a wild boar, and they carry it low towards the ground. They take delight in muddy pools, and are filthy in their habits. They are not of that description of animals which suffer themselves to be taken by maidens as our people suppose, but are quite of a contrary nature (218).

Self-evidently we are dealing here with first impressions written up long after the event and presented with an admixture of poetic licence. By no means all of what Marco records would be regarded as accurate today. Nevertheless he clearly *has* observed a real rhinoceros, has recorded his impressions, and has reflected on these in the light of what he knows of the relevant tradition – that the unicorn (often referred to as *rhinoceros*), cannot be tamed by anyone except a pure virgin, whom it approaches and in whose lap it gently lays its head. For all its ham-fistedness, Marco's rational observation of nature has enabled him and his reader to move on, paradigmatically one might say, from a conception of the *rhinoceros* as a mythical symbol of Christ and Mary to one which involves a real animal of interest in its own right.

V

In the thirteenth century, then, knowledge about the natural world was growing, primarily on the basis of a combination of empirical observation and critical reflection. That very fact, however, seems to have engendered a perceived need to arrange, to systematize that knowledge in order that it might be properly understood and effectively used. This too is a sign of that century's increasing Aristotelianism.

Aristotle himself had before him a huge amount of information about the natural world that was waiting to be sifted, arranged and systematized; and, in the *Historia animalium,* he sought to do precisely this. He described approximately 500 species and divided them into a total of ten groups: human beings, viviparous mammals, egg-laying mammals, birds, fish, cetaceans, soft-shelled molluscs, hard-shelled molluscs, crustaceans and insects. Over the course of nine books he proceeded to discuss these species first in morphological and then in ecological terms – that is to say, Books I–IV focus mainly on forms and body parts, and Books V–IX mainly on habits and behaviour (including, at some length, reproduction). Within his work's subdivisions, one can observe also a strong tendency on Aristotle's part to move systematically between the general and the specific. As a rule, his discussion of a particular physical or behavioural feature begins by concentrating on those aspects of it which all creatures have in common, before moving on to genus- or species-specific differences and inconsistencies.

That said, my use in the foregoing paragraph of terms such as "mainly", "tendency" and "as a rule" was deliberate and programmatic. There is no doubt that Aristotle was a clear, disciplined and systematic thinker; but his zoological works in particular do not always reflect this. The reader of Book VIII of the *Historia animalium,* for example, will find him seemingly almost flitting between such diverse subjects as animal psychology, the effects of climate on behaviour, feeding habits, migration and hibernation. The reasons for this apparent discrepancy between rigidly organized theory and sometimes episodic or aporetic practice are not easy to discern.

On one level, for sure, Aristotle was perennially overstretched and short of time; on another, though, Jonathan Barnes (62f.) is doubtless right to point to his position at the very beginning of a science which he had, to all intents and purposes, founded:

> Aristotle's system is a design for finished or completed sciences... The sciences which Aristotle knew and to which he contributed were not complete, nor did he take them to be... Aristotle says enough to enable us to see how, in a perfect world, he would have presented and organized the scientific knowledge which he had industriously amassed. But his systematic plans are plans for a completed science, and he himself did not live long enough to discover everything.

One sometimes senses that, perhaps for reasons such as this, thirteenth-century recipients of Aristotle saw a responsibility to take the process of systematization he had adumbrated a stage further – encouraged in this also, one assumes, by the marked classifying and standardizing propensities of their own age. As we have seen, Frederick II's *De arte venandi* is far more than a manual of practical falconry. And it is this not least because of the Emperor's extremely careful structuring of his material, which emerges in some ways as more Aristotelian than Aristotle. The work as a whole is punctiliously organized into six books and some 260 short chapters, each with a clear heading and theme. Five of the six books deal with the theory and practice of falconry; but these are preceded by a book which offers a systematic synthesis of knowledge about birds in general, dealing over the course of 57 chapters with their habitats, nutrition, migratory habits, reproduction, form, plumage, flight and defence mechanisms. The overall layout of Frederick's volume, then, already evinces an indebtedness to Aristotle's method of beginning with the general and then moving to the specific; but this principle applies within books and sections of books also. In Book I, for example, Frederick initially divides the avian kingdom into the three categories of waterfowl, land birds and so-called "neutral" birds ("those that may change from one habitat to another", 7), before discussing characteristics of each class in turn. Later, he initially introduces the subject of "migration to escape the cold" in a general way, before moving on to consider in detail which species migrate at which season, why they do so, how they prepare for their journey, where they start out and where rest, on what basis they choose their winter quarters, and how they go about their return flight. Finally, in the second half of Book I, his meticulous description of the individual parts of birds' bodies (from eyes to testes and ovaries) is preceded by an appropriately general chapter on "the functions of avian organs".

If anyone could out-Aristotle Aristotle, however, not least when it came to the systematic organization of a vast amount of material, that person was Albertus Magnus. The structure of his *De animalibus* is a complex amalgam of the ordering principles of his two main sources, Aristotle and Thomas of Cantimpré, with the addition of some – very telling – elements of his own. The first 21 of Albert's 26 books follow the overall structure of the Aristotelian corpus as made available to him by Michael Scot – on which, after all, the *De animalibus* is primarily designed as a commentary. Into this material, however, Albert introduces a great many divisions and subdivisions. With the exception of XIX and XXI, each book is divided into anything between two and six tracts or treatises (*tractatus*), and each of these is in turn divided into a number of chapters. There are in all no fewer than 361 of these, and all have their own pragmatically chosen title. Procedures such as this may seem pedantic, even obsessive; but in fact they are enormously helpful for any reader wanting to find her way around the thought of either Aristotle, Albert, or both. As we have seen, the former's organization of his material can in practice be loose, and he provides no titles or subtitles. Hence one could certainly argue that Albert's meticulous ordering and signposting constituted in itself a significant boost to the reception of Aristotle.

The structure of Albert's Books XXII to XXVI, meanwhile, owes little to Aristotle and much to Thomas of Cantimpré. In essence Albert reproduces, albeit with many changes and additions, Thomas's Books IV to IX (on quadrupeds, birds, sea monsters, fish, snakes and vermin; Albert merges the third and fourth of these into one). Given that, within these books, he preserves Thomas's principle of alphabetical ordering by animal name, it might reasonably be thought that Albert has jettisoned Aristotle's "general to specific" ordering principle entirely. In fact, however, one could argue that he has ensured its at least partial retention. On the level of *De animalibus* as a whole, the addition of five catalogues of individual species following 21 books of often more general Aristotelian material could certainly be seen as reflecting such a progression. Moreover at the head of each individual book from XXII to XXVI, Albert inserts what amounts to a prologue, followed by some relatively brief remarks of a general nature. His words at the very beginning of Book XXIII, for example, reveal particularly plainly his desire to persuade his readers (and himself?) that its structural principles are of a piece with what has gone before, and as such continue to employ both scientific and philosophical (for which two words read "Aristotelian") methodology:

> In this book the nature of birds will be treated specifically, and since every scientific investigation moves from the general to the particular, we will first speak in general about the nature of birds. Afterward, moving according to the order of the Latin alphabet, the birds will be set forth by name in accordance with their species and types. Though it is granted that this procedure is not entirely philosophical insofar as in it the same thing is repeated many times because one and the same thing may pertain to many birds, it nevertheless is an effective procedure for easy teaching and many of the philosophers have held to this procedure. Since, however, we have treated the generation, food, habits, members, and eggs of birds in a general way in our previous books, we have to speak here only of those things in which birds correspond to other animals and those in which they differ from them (1544).

VI

Hitherto in our survey of Aristotle's influence on the thirteenth century we have not said much about his, or indeed Albert the Great's, specifically philosophical enquiries. Yet both were very much philosophers as well as naturalists, and it is to a particular key aspect of Aristotle's philosophical legacy that we must now finally turn.

This aspect is his analysis of the so-called anthropological difference between human and non-human animals. This, after all – especially but not only after its development by René Descartes – remained very much part of the Western philosophical mainstream well into the modern period. Aristotle's point of departure was the notion that there exist various different kinds, or levels, of *psuchê* – an ultimately untranslatable term that tends to be rendered "soul", but also conveys something of the idea of "life force" or, to use Barnes's term (105), "animator". Plants possess simply a nutritive or vegetative *psuchê*, which as such facilitates only growth, feeding and reproduction. Animals have, in addition to this, a sensitive or perceptive *psuchê*, which enables such things as movement, sensory perception and, by extension, desire. Only humans, however, possess a rational *psuchê*, which makes possible thought and the exercise of will (cf. Wild 49). Hence it is only humans who need, and possess, the power of speech. The passage in the *Historia animalium* which goes closest to explaining all this is the following one, from Book II:

Some things possess all the powers of the soul, others some of them, others one only. The powers we mentioned were those of nutrition, of perception, of appetition, of change in place, of thought. Plants possess only the nutritive power. Other things possess both that and the power of perception. And if the power of perception, then that of appetition too. For appetition consists of desire, inclination, and wish; all animals possess at least one of the senses, namely touch; everything which has perception also experiences pleasure and pain, the pleasant and the painful; and everything which experiences those also possesses desire (for desire is appetition for the pleasant)... Some things possess in addition to these the power of locomotion; and others also possess the power of thought and intelligence (quoted from Barnes, 105f.).

Such differentiations between the human and the animal were to prove highly influential in the centuries following Aristotle's rediscovery in early thirteenth-century Europe. Whilst they exclude the possibility, beloved of St Augustine and the Neo-Platonists, of the body and the soul leading a separate and qualitatively different existence, his ideas are not fundamentally in conflict with, say, the first chapters of Genesis or the Church Fathers' exegesis of these; and they were adopted more or less wholesale by the Albertus Magnus pupil Thomas Aquinas in his authoritative *Summa Theologica* (c. 1270).

A decade or so before St Thomas, we see Albertus Magnus both taking on board and struggling with these Aristotelian conceptions of the anthropological difference. Passages like the following make it abundantly clear that he shares Aristotle's basic sense of human superiority, and of this superiority being evident particularly in matters of rationality and the intellect. Albert also conveys, like Aristotle, a sense of a close-knit and active continuum existing between body and soul; and his Christian faith leads him, inevitably, to perceive in humankind a unique aspect of the divine. None of these things prevent him, however, from injecting a certain sting into the tail of his otherwise largely predictable discussion:

For the passions of the powers of the soul are found in many animals. But the differences among these passions in the human are quite evident, for that soul lives most perfectly which has the noblest and most varied powers. For it has the divine, the animal, and the intellectual powers and it impresses its passions onto the nature of the body. Sometimes, a passion that is caused by the body flows back into the soul. Therefore, the difference among these passions in the human is evident. I am calling passions things like fear, boldness, anger, and sexual desire, and in these the rest of the animals share. In some of them, however, a certain power is also found which is like the cogitative and compositive (that is, collative) power in the human (586).

Albert goes on to state that this "estimative power" perceptible in some animals (he gives the example of bees) is given to them specifically to enable them to perform certain activities; but he also suggests that, in these and other respects, there is no difference in the use of the intellect between the child's soul and that of the "brute animal" (587). This latter statement in particular betokens a slightly troubled awareness on Albert's part that, for all its compelling clarity, Aristotle's assessment of the human and animal *psuchê* may not always be able to do full justice to the complexities and fluid borderlines of the natural world. Elsewhere, for example, he evinces a clear belief that nature makes it her business, as it were across the board, to provide appropriate spectra or continua: "Nature never causes genuses to be distant without creating some medium between them, since nature only passes from extreme to extreme through a medium" (287). And, when it comes to the difference specifically between man and animals, he seeks to square the Aristotelian circle by assigning to a particular genus, pygmies, a kind of intermediate, transitional status. His most concerted treatment of pygmies comes in Book XXI:

> Some [animals] flourish so much in the instruction of hearing that they even seem to signify their intentions to one another, as does the pygmy, which speaks, although it is an irrational animal nevertheless. For this reason the pygmy seems to be the most perfect animal, in terms of the animal virtues, after the human. And it seems that among all the animals, it preserves its memories so much and perceives so much from audible signs that it seems to have something imitating reason, but it lacks reason... And thus the pygmy, although it speaks, nevertheless does not argue or speak of the universals of things, but rather their voices are directed at the particular things of which they speak. Its speech is caused by a shadow echoing in a defect of reason... The pygmy is midway between the human, who has a divine intellect, and the other mute animals, in whom nothing of the divine light is perceived to exist, insofar as it uses experiential cognition through a shadow of reason, which it alone of the other animals receives. Nevertheless, with respect to nature it is nearer the brute than the human (1416–18).

Such a passage could hardly be said to read well today – though for certain Albert had no clear idea of what a pygmy really was, and will never have encountered one. Even if one leaves aside modern concerns about taste and respect for human dignity, however, there is no denying that his discourse comes across as both confused and confusing. The pygmy is "the most perfect animal... after the human", "midway between the human... and the other mute animals", yet also "nearer the brute than the human". It is able to speak, to remember, to respond to audible signals in ways that imply the possession of something at least closely akin to human powers of reason; and by implication it might even reflect something of "the divine light" which is denied all other non-human species. Nevertheless, Albert is at tortuous pains to assert that there *is* some kind of intrinsic qualitative distinction between pygmies and humans.

None of Albert's assessment of pygmies constitutes good science, and in its vagueness and implicit prejudice it hardly enhances his reputation as a forward-looking, enlightened scholar. Yet it is also fascinatingly typical of the thirteenth century – in the sense that it evinces a marked and some-times tortured uncertainty about what one might call the real-life border-lines between the human and the animal.

As we will see especially in Chapter 4, these questions were explored in various ways in medieval imaginative literature, primarily through the prism of relationships between knights and their horses. For now, in the context of a chapter centring on Aristotle, we ought to note that, in the thirteenth century, the great philosopher himself was used as a case study of the mutual permeability, indeed interchangeability, of human and equine identity – albeit in the context of two predominantly comic short texts, the French *Lai d'Aristote,* now attributed to Henri de Valenciennes and dated around 1215, and the German *Aristoteles und Phyllis,* which could have been written at any time between 1260 and 1287.

In his capacity as tutor to Alexander the Great, Aristotle has cause to chastise the King for having succumbed to the power of love, to the extent of neglecting his duty by spending all his time with his mistress (in the German version called Phyllis). This advice works only in the very short term; before long Alexander and Phyllis are again in amorous contact, and vowing to get even with Aristotle. Their plan is put into action the next morning, when a scantily clad Phyllis begins to disport herself seductively in front of Aristotle's study, and eventually inflames his passion to such a degree that he leaves his books, grabs the maiden by the tunic and expresses a strong desire for carnal knowledge of her. In the German version he also offers her money (*Novellistik* 386–9). She agrees to "play the wanton" (Henri 340), on one condition:

"You must do a very singular thing for me, if you are so infatuated with me. For I have conceived a very great yearning to ride on you a little, on the grass in this fine garden, and I would also like", said the maiden, "to have a saddle put on you, so that I shall ride with greater honour." The tutor replied joyfully that he would do this gladly, as someone entirely devoted to her. Nature had certainly disturbed him, when the maiden made him carry on his shoulders a palfrey's saddle. Love makes a fool of a wise man if Nature urges him on, when the very finest scholar in the world has himself saddled like a packhorse, and is then made to go on all fours, crawling over the grass (Henri, trans. Brook/Burgess, 341–61).

In the German *Aristoteles und Phyllis* the dehumanization involved is intensified by the fact that Phyllis removes her belt and uses it as a bridle (407–9). In both texts, though, the message is abundantly clear: sexual desire is so strong that it can overpower even the most superior intellect, and swiftly reduce even the most distinguished human being to the level of a low-grade animal – it is for certain no coincidence that Aristotle is here associated with a palfrey or packhorse, rather than with, say, a knight's destrier. Similarly, nature is stronger than culture, so much so that it can easily turn topsy-turvy the normal expectations of the civilized world – as Brook and Burgess aptly put it (34), "metaphorically, it is Aristotle, whose aspirations of sexual fulfilment are about to be shattered, not the maiden, who is being taken for a ride".

There is of course no reason to assume that either the French or the German Phyllis poet had read Aristotle, or indeed any of the Latin-language learned literature his thirteenth-century rediscovery so impressively spawned. Nevertheless they, and presumably their audiences, clearly had the idea that Aristotle was somehow the best of all scholars; and, for all their focus on entertainment, they were plainly also fascinated at some level by the complexities of the relationship between what is human and what is animal. Even in their light-hearted and hardly flattering way, therefore, these poems reflect the fact that Aristotle's rediscovery was both significant and opportune.

More generally there can be no doubt that, in thirteenth-century Europe, Aristotle's work on nature fell on uncommonly fruitful soil. As always in the history of thought, the role of the chicken is impossible to differentiate clearly from that of the egg – was Aristotle the first to ignite a new way of looking at animals in the thirteenth century, or was his contribution more by way of a potent fanning of already slow-burning, perhaps subliminal flames? What is certain is that, by 1300, the scientific study of animals along Aristotelian lines had greatly advanced, and habits of empirical observation and analysis had become far more deeply ingrained than they had been in 1200. In this respect there had been an Animal Turn. Our next chapter will demonstrate, however, that such developments in no way killed off earlier medieval conventions of attaching spiritual interpretations to real or imagined animal behaviour in order to instruct, rebuke or challenge Christian believers. Thomas of Chobham stated, after all, that "there is no creature in which we may not contemplate some property belonging to it which may lead us to imitate God, or some property which may move us to flee from the Devil"; and he too was a thirteenth-century figure. Far from working against it, indeed, the integration of Aristotelian thought into mainstream European intellectual culture can be seen significantly to have promoted and expanded – and also grounded more firmly in experiential reality – the use of animals to convey moral or religious meanings, especially to the laity. It could not have done this in the absence of other far-reaching initiatives and developments that took place in the early decades of the thirteenth century; and it is to thes e that we must now turn.

CHAPTER 3

Innocent III and Thirteenth-Century Animal Imagery

Abstract This chapter studies the use of animals to convey ethical and theological meanings to non-clerical audiences. It shows that such initiatives, whilst far from new, were given added impetus in the thirteenth century by the fruitful combination of two superficially divergent trends – a new papal and clerical determination to improve the theological education and moral behaviour of the laity, not least through sermons; and the "new" Aristotelian ways of looking at the natural world. These provided preachers and pastors with fresh material about animals and new ways of interpreting this allegorically and of organizing it into user-friendly compendia. They also breathed new life into traditional animal interpretations from the *Physiologus* and elsewhere, particularly by basing them more plausibly on the behaviour of "real-life" animals.

Keywords Allegory • Aristotle • Innocent III • *Physiologus* • Preaching

I

Innocent III (Pope from 1198 to 1216) was one of the most powerful and important of all medieval holders of that office. He played a profoundly interventionist role in the politics of Europe, above all those of the Holy Roman Empire; he introduced far-reaching reforms in the field of canon law; and he was a prime mover in the Fourth Crusade of 1202–4. Prior to

© The Author(s) 2020
N. Harris, *The Thirteenth-Century Animal Turn*,
https://doi.org/10.1007/978-3-030-50661-2_3

his assumption of the papacy, he was also a prolific author on moral theological matters, noted especially for his distinguished contribution to the *contemptus mundi* tradition, *On the Misery of the Human Condition.*

In addition to all this, Innocent can be seen to have made a major if indirect contribution to changing perceptions of animals in thirteenth-century Europe. This is because of his summoning and prosecution of the spectacularly well-attended Fourth Lateran Council, which took place in Rome in November 1215. This Council was consumed above all by two related preoccupations: the need to improve the Church's frequently abject – indeed non-existent – pastoral and theological ministry to its laity; and the need to extirpate heresy, particularly that of the ascetically dualist Cathars. Strange though it might initially seem, the processes set in train by Innocent to wage war on these two fronts were, over the coming decades and centuries, to make extensive and effective use of animals.

With regard to heresy, the Council increased the scope and powers of the Inquisition and promulgated three canons (or decrees) aimed specifically at countering the Cathar threat. These defended the Church in the face of radical questioning, materially sharpened punishments to be administered to heretics, and above all programmatically restated relevant historic doctrines on the Trinity, and on the incarnation, death and resurrection of Christ. Meanwhile the key canon aimed at promoting lay piety was the seemingly unremarkable 21st, headed "On confession being made, and not revealed by the priest, and on communicating at least at Easter". This imposed on all who had reached the age of discernment the dual requirement to engage in auricular confession and to receive the Eucharist at least annually – not, one might think, a particularly onerous demand, but one which had never been formally stipulated before. Thanks no doubt to Innocent's measures to improve clergy training and streamline administrative processes, this canon proved remarkably successful, to the extent indeed that it is still widely respected and followed in Catholic circles today. Moreover this attempt on his part to strengthen the Church's ministry of the Sacrament was accompanied by a comparable determination to expand and deepen its ministry of the Word. Innocent himself was an assiduous preacher; already by the end of the twelfth century sermons had become increasingly common, at least in learned Latinate circles; and in the thirteenth and fourteenth centuries they were to grow enormously in number and in liturgical, theological and pastoral significance, above all in the context of ministry to the laity.

One of Innocent III's last acts before his death in 1216 was to issue a Bull approving the Dominican Order; and in the following decades it was to be above all this mendicant body that carried forward the cause of preaching as a means of instructing and improving the laity. Seldom has there been a better case of *nomen est omen*. Homiletic activity is, after all, fundamental to the Dominicans' very title, the *Ordo Praedicatorum;* and the much cited popular etymology of *Domini canes*, "dogs of the Lord", further implies how ideally suited the order was to carrying forward the agenda set out at the Fourth Lateran Council. From patristic times the perceived healing properties of dogs' tongues in particular had been associated with the beneficial effects both of preaching and of the sacrament of penance: the late thirteenth-century poet Brun von Schönebeck expresses this with admirable economy in his paraphrase of the Song of Songs, when he states that "my tongue says that they [good priests] are God's dogs – their tongues heal wounds with confession, preaching and teaching" (11574–7; Schumacher 38). Meanwhile a dog's bark can be an effective image of the rather different kind of utterance a preacher or inquisitor might employ towards reprobates or heretics: St Dominic's successor and first biographer Jordanus of Saxony (d. 1237) already recognized this, by referring to Dominic's ability to "awaken souls sleeping in sin by the barking of holy instruction" (Schumacher 50).

Even this straightforward canine example demonstrates how useful animals can be when conveying an idea, or a moral or theological lesson – especially, perhaps, to a lay audience unfamiliar with the abstractions and complexities beloved of medieval intellectuals. This insight was by no means new in the thirteenth century: on the contrary, it is intrinsic to the traditional concept of nature as a book which God uses to communicate with man. It is important to note, however, that the "new" approach to animals evinced by Aristotle and his thirteenth-century followers was to prove in no way incompatible with older didactic traditions such as this. Indeed, it helped to give them something of a renewed impetus – and in ways that are often strikingly reminiscent of the ways we have seen Aristotle influencing thirteenth-century scientists and encyclopaedists. First of all, the new impulses in animal studies seem to have resulted in animals being "in", at least in clerical circles; and so the thirteenth century – as well as the first half of the fourteenth – witnessed, in purely quantitative terms, a marked increase in the frequency with which they were used to reinforce the kind of lesson Innocent III, the Dominicans and many other like-minded figures felt that lay Christians needed to learn. Secondly, in response to this trend and to the passion for organization and systematization we observed above in writers such as Albert the Great, a number of predominantly mendicant authors began to compile significant collections of exempla designed specifically for use by preachers – a process which culminated, between roughly 1280 and 1340, in the appearance of three massive specialist compilations of nature exempla. Thirdly, thanks to the assiduity of the thirteenth-century encyclopaedists, especially Thomas of Cantimpré, a number of "new" animals were introduced to enable authors and preachers to convey new spiritual meanings – be they obscure creatures known hitherto only to scientists or encyclopaedists or, at least in one case, a very familiar one which became deeply embroiled in the struggle between orthodoxy and heresy. Fourthly, the characteristically thirteenth-century emphasis on observing animals qua animals led, in certain cases, to their symbolic use taking on a more realistic, naturalistic character. And finally, some long-established, "canonical" animal allegories, such as those found in the *Physiologus*, were subjected to innovative treatment, often resulting in their taking on unpredictable new meanings and literary forms. In these respects at least, the venerable Book of Nature itself experienced an Animal Turn.

II

The extent to which animals became fashionable can be seen – as one might expect – not least in sermons. After centuries of relative neglect, preaching had begun to flourish again in the twelfth century, if only in clerical contexts; but the use of animals to illustrate or underline theological content was very much a phenomenon of the post-Lateran thirteenth and fourteenth centuries. The practice burgeoned particularly, though far from exclusively, in the works of mendicant preachers, above all Dominicans – who, as we have seen, were closely involved with the mission to instruct the laity initiated by Innocent III, but were also frequently associated with the university milieux in which the works of Aristotle and his successors were at their most widely available and influential. It is no coincidence, for example, that Albert the Great, Vincent of Beauvais and indeed Thomas Aquinas were all Dominicans who attended the University of Paris.

Limiting ourselves just to the thirteenth century, we can see – thanks especially to the valuable spadework done by Elisabeth Schinagl (2001) – that animals featured prominently in, for example, the sermons composed by the Prague Dominican Martin of Opava and by the anonymous late thirteenth-century Dominican whose sermons are wrongly attributed to Albert the Great in the Leipzig manuscript UB 683. Martin of Opava (also known as Martinus Polonus) compiled, some time before his death in 1278, a set of 321 model *Sermones de tempore et de sanctis* aimed at preachers active in an urban lay setting – it was common practice for such materials to be written down in Latin, even if they were eventually intended for oral use in vernacular sermons. Martin's homilies make exhaustive use of narrative exempla drawn from biblical, classical and more modern historical sources, but also contain numerous spiritual interpretations of animals. Some of these are entirely conventional: both the pelican (Sermon 283) and the phoenix (51) are compared respectively to Christ's sacrificial death on behalf of his children and resurrection on the third day, much as they had been innumerable times since the early Christian centuries. At other times much retailed *Physiologus* stories occur in only slightly modified form, as when the temptations of the world are compared to the whale, who "emits sweet breath from its mouth, to breathe in which the fish come together, whereupon the whale ensnares and devours them" (103). Elsewhere, however, Martin treats well-established allegorical traditions with rather greater freedom: he presents the mole as an emblem of cupidity, though not for the usual reason that it is blind, but rather because "it always digs in the earth and cannot live anywhere other than in the earth" (171). Moreover the bird he uses to exemplify filial piety is not, as one might expect, the eagle or the stork, but rather the crane (28); and there are at least a few occasions on which he shows a considerable degree of independence, even originality. He takes up a *proprietas* of the raven, found in Aristotle and in the thirteenth-century encyclopaedias, to the effect that the bird, "when it attacks another bird, first of all puts out its eyes, so that, blinded, it cannot escape"; this is then compared to the blinding properties of sin (123). Finally we see Martin making what is almost certainly a unique, if possibly garbled association between Christ and the king bee (*sic*): "the king of the bees is by nature noble and distinguished, such that it exhibits magnitude and beauty, as well as what is most important in a king, gentleness. For even though it has a sting, it does not use it in judgement" (4). Another such king, of course, is Christ.

The animal images used in Leipzig MS 683, meanwhile, are remarkable less for their originality (or indeed length – they are nearly all very short) than for their sheer number. Schinagl's survey (156–9) mentions five references to the elephant, four to the bee, three to the bear and two each to the eagle and to a curious sea creature called the *monachus marinus*, as well as brief comparisons or allegories involving the panther and lion (their *Physiologus* characteristics), crocodile, basilisk, mole, bat, ape, sheep, dolphin, toad, hen and stag. It is clear that even by this stage the repertoire of animals used to convey moral and/or spiritual meanings was expanding considerably; and this process was to continue into the fourteenth century, reaching its apogee, at least with regard to sermons composed in Germany, with the frequent and often detailed animal moralizations in the *Sermones Socci* of the Cistercian Konrad von Brundelsheim (d. 1321).

III

This increase in the range of animals known to scholars was accompanied, as was the case also with the reception of Aristotle's zoological works, by moves to arrange and systematize that knowledge, with a view to making it more readily accessible. Among the many "new instruments of intellectual work" (Berlioz/Polo de Beaulieu 181) which the thirteenth century produced in abundance – encyclopaedias, *distinctiones*, repertoria, indices, biblical concordances – can be numbered several collections of exempla designed primarily for the use of preachers. In the earlier thirteenth century stories concerning animals or reports of their natural characteristics occurred relatively seldom in these: animals hardly appear at all, for example, in Caesarius von Heisterbach's *Dialogus miraculorum* (1219–23), which is cast in the form of a dialogue between a monk and his novice. A turning point of sorts came shortly after the arrival on the scene of the widely used encyclopaedias of Bartholomaeus Anglicus and Thomas of Cantimpré. Between 1250 and 1261 the Dominican inquisitor Stephen of Bourbon compiled his massive, yet incomplete *Tractatus de diversis materiis predicabilibus*, which contains nearly 3000 exempla, approximately a sixth of which are either devoted to or prominently feature animals. The exempla are ordered in seven books according to the Seven Gifts of the Holy Spirit: fear, piety, knowledge, strength, good counsel, intelligence and wisdom (only the first four and a half books survive). Each of these books has clear subdivisions; but within these, the individual exempla are not arranged in an obviously systematic way. It is true that those involving animals sometimes come in clusters – over the course of ten exempla in Book I, Chapter 1, for example, there are two moralizations of the lion, one each of the peacock and stag and an anecdote about the monk

Theodore and a dragon; but the exempla between and around these are related to them only in being generally about the spiritual gift of fear. Hence the animal material can have been easy to locate only in the context of a manuscript with an unusually full and reliable index.

This kind of practical problem is no doubt one of the reasons for the development, in the later thirteenth century, of specialist collections of nature exempla, based in the main on the *De proprietatibus rerum* of Bartholomaeus Anglicus. The first of these chronologically seems to have been the *Liber de moralitatibus* (or *Tractatus septiformis* or *Proprietates rerum moralizate*) attributed in two of its seventeen manuscripts to one Marcus of Orvieto, but manifestly of Franciscan origin and datable to the 1280s (Etzkorn 187f.). As its various titles suggest, Marcus's volume is devoted entirely to moralizations of all kinds of natural phenomena and organized in seven *tractatus*: on the heavenly bodies, the elements, birds, fish, animals, trees and plants, and precious stones. Within these sections – whose number and distribution have obviously been inspired by the seven days of creation – the individual species or phenomena are dealt with alphabetically, in discrete chapters. Nearly all the individual moralizations also follow the same basically tripartite structure – the natural historical *proprietas* is followed by the spiritual interpretation and then (usually) by a series of quotations from authorities.

Much the same approach is followed by the enormous *Reductorium morale* of the Benedictine Petrus Berchorius (or Pierre Bersuire), which is heavily dependent on Marcus's work but was compiled only in the years between 1325 and 1341. The *Reductorium* differs fundamentally from the *Liber de moralitatibus* only in its overall structure and in its sheer size. It is in sixteen books, the first thirteen of which follow faithfully enough the layout of Bartholomaeus's encyclopaedia. Then, however, come three books of an entirely different character – a book "on the miracles of the natural world", an *Ovidius moralizatus* and, finally, a commentary on selected biblical passages entitled *Super totam bibliam*. All sixteen of the books, however, reveal a quite monumental conscientiousness, concern for detail and indeed inventiveness. Marcus of Orvieto's collection is not exactly a slim volume (its modern critical edition occupies exactly 1000 pages); but it is frankly dwarfed by that of Berchorius. A random selection of four chapters at the beginning of the two works' sections on animals reveals, for example, that Marcus has four moralizations of the wild boar *(aper)*, ten of the bee *(apis)*, six of the spider *(aranea)* and nine of the ass *(asinus)*; the equivalent figures for Petrus are 15, 88, 31 and 26 respectively. Whether because or in spite of that compendiousness, the *Reductorium morale* proved successful in the Middle Ages and, not least, the early modern period (seven editions between 1521 and 1731). Certainly both it and Marcus's work are very easy to use for anyone interested in spiritual interpretations of animals, even without a detailed index.

What, though, of the preacher who might at times prefer to look up a potential topic to preach on first, rather than beginning his search with the name of an animal? Such a question may well have been posed by the compiler of the third great turn-of-the-century nature exemplum collection, John of San Gimignano (or Johannes Gorus). His *Liber de exemplis et similitudinibus rerum*, from around 1300, seeks, in terms of user-friendliness, to get the best of both worlds by steering a kind of middle course between the approaches of Stephen of Bourbon on the one hand, and Marcus and Petrus on the other. As Christel Meier puts it (118f.), John "makes use of two conceptions of the classification of knowledge. While he arranges the different books of his encyclopaedia according to a systematic order of things *(significantia)*, he makes use of an alphabetical order of meaning *(significata)* in each of these books". For example John's Book IV is headed "De natatilibus *(sic)* et volatilibus", and hence is mainly concerned with fish and birds; but its seventh to tenth chapters are headed respectively "Flatterers or heretics" *(Adulatores seu heretici)*, "Friend" *(Amicus)*, "Avarice" *(Avaritia)*, "Blasphemers" *(Blasphemantes)* and "Good and virtuous men" *(Boni et virtuosi viri)*. One has to read the main body of the respective chapters – where the moralizations again follow the classic tripartite scheme – to realize that the animals being interpreted include the eagle, the sparrowhawk and other birds of prey, fish, sea monsters and bees. Whether this approach – which John maintains throughout his ten books – actually represented the worst, rather than the best of both worlds presumably depended on the needs of the user, and on the presence or absence of an index.

IV

That such estimable works of reference as the three we have been discussing were needed in the first place is due to the thirteenth-century popularity both of animals and of sermons, and to the number of new species which were beginning to feature in moral theological contexts for the first time. Knowledge of these came from various sources, as venerable as Aristotle or as recent as Jacques de Vitry; but it was materially promoted above all by the *De natura rerum* of Thomas of Cantimpré. This was partly because Thomas scoured earlier nature encyclopaedias and comparable compendia more assiduously than anyone had before him, and partly because he himself provided spiritual interpretations of many animals he wrote about. Baudouin van den Abeele (1999b) has calculated that, of the over 500 animals Thomas discusses, he includes at least short allegorical moralizations of some 116 – in line, of course, with the desire expressed in his prologue to further "the proclamation of the faith and the correction of morals" (5). It is doubtless revealing in this context that Thomas does not interpret many of those animals familiar from the *Physiologus* and other older traditions – such as the hedgehog, weasel, fox, lion, beaver or hyena (van den Abeele 1999b, 131). Rather, for all his Aristotelian credentials, he seems to have had the clear intention of making available preachable material about numerous creatures with which his readers will have been unfamiliar – and, indeed, whose natural historical characteristics they might well have found implausible. This trend is at its most marked in Thomas's book on fish. For example, we learn in quick succession about the *gradus* (VII, 41), the *irundo maris* ("sea swallow" – VII, 42) and the *mugilo* (mullet – VII, 50). The first of these has a single eye at the top of its head, and hence can represent those contemplatives who always keep their eyes fixed on God. The sea swallow, for its part, is a curious mixture of fish and bird, which lives in the sea, yet has wings which it can use to fly into the air when required. As such, it can be interpreted as those who cannot avoid becoming embroiled in the affairs of this world, yet are able at the right time to rise above these and contemplate spiritual matters. Finally, the *mugilo* believes that, if it buries its head, the whole of its body will remain safe. This is reminiscent, Thomas argues, of those foolish people who believe that, if they guard their heart, their entire body will be safe from worldly dangers.

It could not be said that Thomas's moralizations of these three fish inaugurated an extensive career for any of them in religious writings; in a few cases, however, some recondite creatures whose characteristics were interpreted by Thomas did find their way into a respectable number of later texts. One of these is the unidentifiable small animal he calls *leoncophona*, which is deadly to the lion. When captured by hunters, it is burnt to ashes, and these ashes are scattered in the path of lions, who are killed if they but touch them. A similar fate awaits any lion who either bites a *leoncophona* or comes into contact with its urine – which, accordingly, the animal emits proactively if it knows a lion is coming (IV, 73). For Thomas, the *leoncophona* represents humble souls who, burning with the love of God, spread it before the proud and hence defeat them.

By the 1250s knowledge of this anecdote had clearly reached Stephen of Bourbon, who lists the *leontofonus (sic)* as one of the seven things a lion fears; in the same way, he suggests, men should fear the poison of sin (I, 1, 374–85). In the course of its subsequent treatment by later authors also, the *leoncophona* tends to veer between positive and negative interpretations, depending on the meaning attributed to the lion in the context in question. Konrad von Megenberg and one interpretation by John of San Gimignano repeat Thomas's association between the animal and humility; and the various versions of the *Etymachia* treatise use it as an emblem of patience. By contrast, in two interpretations by Petrus Berchorius the lion represents a good man, so the *leoncophona* becomes either the Devil or an evil temptress; and in a second, contrasting moralization by John of San Gimignano it means – as for Stephen of Bourbon – sin itself, which holy men (lions) rightly fear (see Harris 1994, 388–90).

As these and other instances show, new species introduced into moral theological discourse in the thirteenth century tended to be little known, sometimes bizarrely implausible and/or rather exotic – in the case of the *leoncophona*, indeed, one suspects that no such animal ever existed outside stories concerning the lion, and that its name originally meant "lion-killer" (with a Greek *-chtone* ending). This is hardly the case, though, with the cat, which in the thirteenth century experienced a particularly malignant Animal Turn of its own. Certainly it began to appear much more frequently in theological and imaginative literature than in any previous Christian century, and came to be presented almost exclusively in a highly negative way, representing heresy, a wide range of sins and – perhaps most alarmingly in the light of later developments – the Devil himself.

Although well known in Europe for many centuries as a skilled rodent catcher, the cat is hardly mentioned in the Bible and not at all in the *Physiologus*; and hence one cannot speak of anything approaching a "standard" Christian interpretation of it in the patristic or early-medieval periods. If anything, cats get rather a good press in early saints' lives, where they appear above all as companion animals, not least to monks and nuns. Around 1200, however, the cat fell victim to a calamitously far-reaching etymology suggested by Alan of Lille (c. 1120–1202) in his *De fide catholica*, written at some point between 1185 and 1202: "the name of the Cathars *(Cathari)* comes from cat *(catus)*, because, it is said, they kiss the posterior of a cat – in whose form, it is said, Lucifer appears to them" (I,

83). Alan is probably taking his information about the *osculum infame* from a passage in Walter Map's *De nugis curialum* of 1180; and his attempt at popular etymology was imitated also in vernacular contexts – at least by the distinguished Franciscan preacher Berthold von Regensburg (c. 1210–72), who in his Sermon 45 draws a parallel between the German words for "cat" (*Katze*) and "heretic" (*Ketzer*). The baleful connection between the cat, the Devil, heresy and moral depravity was, however, made most influentially by the notorious Bull *Vox in Rama*, issued by Pope Gregory IX in June 1233 in response to accusations of widespread Luciferianism in northern Germany. Gregory describes some particularly gruesome cultic activity as follows:

> Then all sit down to a banquet and when they rise after it is finished, a black cat emerges from a kind of statue which normally stands in the place where these meetings are held. It is as large as a fair-sized dog, and enters backwards with its tail erect. First the novice kisses its hind parts, then the Master of Ceremonies proceeds to do the same and finally all the others in turn; or rather all those who deserve the honour. The rest, that is those who are not thought worthy of this favour, kiss the Master of Ceremonies. When they have returned to their places they stand in silence for a few minutes with heads turned towards the cat. Then the Master says: "Forgive us." The person standing behind him repeats this and a third adds, "Lord we know it." A fourth person ends the formula by saying, "We shall obey." When this ceremony is over the lights are put out and those present indulge in the most loathsome sensuality, having no regard to sex (quoted from Peters, 157).

There was no way that the cat's reputation could readily recover from authoritative descriptions like this and many others – least of all, perhaps, in the eyes of Dominicans, engaged as they were in front-line combat against heresy and its consequences, and associated as they were with the cat's perceived ancient enemy, the dog *(Domini canes)*. It is for certain no coincidence that an account of the Devil appearing as a cat to St Dominic came to be recounted with some relish by several representatives of the Order of Preachers we have already encountered: by Stephen of Bourbon (I, v, 756–778), by Vincent of Beauvais (*Speculum historiale*, 30, 76) and by Martin of Opava (*Promptuarium exemplorum*, 233).

Nor in this context is it exactly surprising that Thomas of Cantimpré, here receiving and passing on a contemporary interpretative trend rather than seeking to invent one himself, should offer an account of the cat which "reads like a catalogue of vices" (Blaschitz 600). His chapter IV, 76 (151f.), bet-hedgingly entitled "De musio vel murilego, qui etiam et cattus dicitur", does not mention the cat's associations with Lucifer, but does interpret a range of its natural characteristics firmly *ad malam partem*. Cats, Thomas inveighs, are "impure and odious" animals, fight each other "most cruelly and often", and are lascivious, avaricious and lazy. Nor do things seem to have improved by the late thirteenth century: Marcus of Orvieto's chapter on the *murilegu*s (V, 40) contains some eleven spiritual interpretations of the cat, all bar one unequivocally condemnatory. Its acute eyesight, for example, is analogous to that of the "sons of darkness"; its lack of a fixed colour recalls the nature of "hypocrites and apostates"; its large mouth, sharp teeth and long tongue are like those of envious detractors; and various of its behaviours reflect those of identifiable types of sinner – those who fight each other over benefices, those who refuse to acknowledge their misdeeds, those who are more interested in outward appearance than inner virtue and, not least, lascivious women who are captured by the Devil. The disastrous association between (black) cats and women which was part and parcel of the infamous witch-hunts of the early modern period was present only embryonically in the literature of the thirteenth century; but the direction of travel is already only too clear.

The question as to why cats should be demonized in this way is far from easy to answer. Up to a point, as some of our earlier examples have implied, animals could (still in the thirteenth century) be used to transmit any meaning that an author wished to get across. Or maybe, as the etymologies of Alan of Lille and Berthold von Regensburg imply, the cat simply had the wrong name at the wrong time. One suspects, though, that there was more to it than that – especially, after all, in a century like the thirteenth, when even familiar animals were doubtless beginning to be observed more closely and with greater interest than had been the case previously. As Katharine M. Rogers suggests (49), the cat possesses certain "natural abilities [that] can readily be interpreted as evidence of uncanny abilities, which may be perceived as divine or demonic". The cat's silent, yet sublimely precise, movements; its exceptional powers of sight and hearing; its gift of anticipating earthquakes or electrical storms; its fundamental untameability; its skill, indeed cruelty as a hunter; its apparent tendencies towards laziness and vanity – all these and more readily observable traits could easily give the impression that cats had special, perhaps frightening powers, and by no means necessarily in a good way. Aristotelian habits of observation, in other words, could cut both ways – promoting either a more tolerant, progressive view of the world and animals' role in it or the dehumanizing reinforcement of prejudice about certain kinds of non-orthodox opinion.

V

Overall there can be little doubt that, from the thirteenth century onwards, medieval animal imagery became more dependent than previously on realistic, pragmatic appraisals of actual animals. Two case studies should suffice to make the point.

The first of these concerns the peacock. In antiquity and the early Middle Ages the peacock was seen very much as a symbol of resurrection and eternity (see Lother and Schwabe). From the later twelfth century onwards, however, the bird tended to be associated more and more regularly with pride, by authors including – in roughly chronological order – Matthew of Vendôme, Alexander Neckam, Jacques de Vitry, Marcus of Orvieto, John of San Gimignano and Heinrich von Neustadt (Harris 2011, 175). In most cases the *tertium comparationis* cited is the peacock's extravagant tail display, often contrasted with its "greyish horny-brown… legs, which sit oddly with the glamour of the rest of the bird" (C. E. Jackson 22).

It seems entirely reasonable to postulate that this increased frequency of literary associations between the peacock and pride (and for that matter hypocrisy) was a reflection of increasing contact between medieval people and actual peacocks. The latter certainly did become increasingly common in Europe as the Middle Ages progressed, especially though not only in aristocratic circles (see for example Dowling 377). Moreover from 1223 onwards peacocks – generally with raised tails – were in widespread use also as heraldic emblems (H. Hartmann 151, 175). One way and another, then, "peacocks were a common adornment of medieval courts… [and] a more familiar sight to the society of ducal and royal courts… than to any of us [today]" (Wailes 952f.).

A perhaps clearer example of the influence of real-life observation on medieval animal imagery is provided by Stephen L. Wailes's clear-headed attempt to illuminate a textual crux in Walther von der Vogelweide's lyric *Dô Friderîch ûz Osterrîch alsô gewarp* (L 19,29), datable to 1198 or 1199. "Wherever I went, I crept along like a peacock", Walther writes. "I hung my head down on my knees, but now I raise it up in full dignity" (19,32–4). In line with long-held scholarly assumptions, earlier attempts to explain these images sought to do so primarily by searching for literary precedents and authorities – and drew a blank. Wailes shows, however, that if one simply takes the poet's words at face value and in the light of the way peacocks actually behave, the scales soon begin to lift from one's eyes:

> In the course of normal movement, walking or running, the peacock carries its head well above its back. The angle of the neck is perhaps thirty degrees to the vertical. It carries its head low, a few inches off ground, with the neck angled downward, only when it is looking for something to eat. Like other fowl it hunts and pecks, and when it scratches the ground to turn up seeds or insects it drops its head to the ground close to its body. One might easily speak of the peacock's head being down around its knees while the bird is searching for food (953).

From this observation it is but a short step to surmise that the peacock images refer to the poet's own situation vis-à-vis his patrons, and hence his ability to earn a living. When Duke Frederick of Austria died in 1198 (an event alluded to in the poem), Walther suddenly lost his principal source of income. So he was "left to prowl about like the hungry peacock, head to the earth, living (in material terms) from day to day, restless and unsatisfied until established again as comfortably as he had been with Frederick" (Wailes 953). Once, however, he had been taken up – as he was in relatively short order – by Duke Philip of Swabia, Walther could again raise his head well above his knees, like a suitably well-fed peacock.

No doubt an approach grounded in the appearance and behaviour of real-life peacocks can also help explain a significant – but hitherto hardly ever observed – tendency, specifically from the thirteenth century, to associate the bird with senior representatives of the clergy. By no means all of these comparisons were made in a spirit of disapproval. For example Thomas of Cantimpré and, especially, his vernacular adaptor Konrad von Megenberg see the peacock as above all a good bishop. Konrad even manages to provide a positive take on the peacock's ostentatious display of his train: "When people see the bishop's head of justice in his true and pure works, so that his subjects follow him, then he stretches out his train, that is to say, his good works, and evermore draws his subjects into eternal life" (III B 57). On the whole, though, one has the impression that the peacock was more often conscripted into the service of thirteenth- and fourteenth-century scepticism about the effectiveness and probity of the higher clergy, not least the papacy itself. Marcus of Orvieto, for example, includes the following as the fifth of his eight moralizations of the peacock:

> The male follows after the female and searches out her eggs in order to break them, so that he might be free to devote himself more fully to his lusts. Fearing which, the female hides her eggs, so that he might not easily find them. The peacock signifies those prelates whose emotions prevent their subjects from doing good (I, 364).

Marcus does not specifically mention the Pope. In the Latin poem *Pavo* (c. 1285) by the Cologne canon Alexander von Roes, however, the eponymous bird plainly does represent the Supreme Pontiff. Alexander sets a number of avian characters in the context of a General Council of the Church; and there are many hints that the specific Council he has in mind is that of Lyon in 1245, at which Pope Innocent IV notoriously sought to depose Frederick II as Holy Roman Emperor. The papal peacock, with the particularly enthusiastic support of the cockerel – that is, the French King, Louis IX – , summons the (imperial) eagle to his Council, condemns him when he fails to appear, robs him of his wings and feathers, and eventually usurps his power. For all his allusions to events of some forty years earlier, there can be little doubt that the Pope whose behaviour Alexander von Roes wishes here to denounce was not Innocent IV, but rather his own contemporary Martin IV. Martin was French, wholly dependent indeed on the French crown, authoritarian by temperament, and moreover an enemy of Alexander's Rome-based employer in the first half of the 1280s, Cardinal Jacopo Colonna. The increasingly common late-medieval association between the peacock and ecclesiastical dignitaries is, then, here being deployed in the interests of topical political propaganda expressed in indirect, parabolic form – a tradition which continued well into the sixteenth century (see Wirth).

The question is, though, why? What caused authors and artists to forge a link between peacocks, Popes and prelates? The most plausible explanation is surely to postulate that a mental connection is being made between the peacock's tail display on the one hand and opulent clerical robes on the other. The latter had, after all, become increasingly common in the aftermath of the Fourth Lateran Council of 1215. Innocent III sought there to promote the wearing of vestments – not least for the benefit of the laity; and in subsequent decades this practice flourished, especially in the context of processions such as those on the Feast of Corpus Christi, a solemnity which was eventually instituted for the entire Western Church in 1264. Moreover processional vestments – as distinct from standard liturgical ones – came over the thirteenth century not only to be donned more frequently, but also to become ever more opulent (Hayward). This was true in particular of copes, whose large scale, frequently ornate design and basically circular shape bear clear similarities to the raised train of a peacock. Copes worn by bishops and Popes (the so-called *mantum*) tended to be not only particularly conspicuous but also particularly sumptuous; and all in all it does not seem far-fetched to infer that a late-medieval observer, perhaps especially one with an anticlerical and/or puritanical bent, may have been drawn to associate, say, a bishop with a peacock on the basis of a kind of natural symbolism.

VI

In the case of the peacock, then, it really does seem that the thirteenth-century Animal Turn caused its symbolism to be more firmly grounded in the species' actual characteristics than has hitherto been realized. This is equally true of the camel.

Before around 1200, counter-intuitively perhaps, the camel had been primarily a symbol of Christ – on the basis of Gospel verses such as Matthew 19:24, where Jesus says that it easier for a camel (himself) to pass through the eye of a needle than for a rich man to enter the kingdom of God, or alternatively Matthew 23:24, where he denounces the scribes and Pharisees as blind guides, who strain off midges yet gulp down a camel. Gradually, however, this sense of the camel being in essence an emblem of Christ began to change – as people had more and more opportunity to observe and reflect upon what camels were really like. This process no doubt began as early as the First Crusade of 1095–9: there is for example a revealing passage in William of Tyre's *History of Deeds Done Beyond the Sea* (III, 15), where William tells us that, in 1097, the crusaders plundered "many asses and pack horses, [and] troops of camels – animals which our people had never seen before". Following this initial sighting, a growing number of Western pilgrims, as well as crusaders, will have had the opportunity to see camels at first hand; and at least from the thirteenth century onwards, they were occasionally seen in Europe as well, in royal or princely menageries.

In the context of this much greater contact with camels as the Middle Ages progressed, one voice in particular can be identified as having inaugurated a different and in some ways more realistic view of them in European literature. That voice belonged to Jacques de Vitry, the theologian and chronicler who resided in the Holy Land between 1216 and 1225, and whose *Liber orientalis*, part historical account and part record of the author's own experiences, was a significant source for thirteenth-century and later writers of natural history. Jacques brings to bear important information that is not recorded in earlier scientific works, including those of Aristotle. In his chapter 88 he alludes, for example, to the readily apparent fact that the camel appears "misshapen". He adds that camels are lazy, and hence walk slowly; and that, especially when angered, they shriek unpleasantly. Jacques' "new" details, whilst not numerous, plainly derived from real-life observation of camels; and they had the cumulative effect of introducing a greater degree of negativity to evaluations of the camel by Western authors. It is for certain significant that they all found their way into the highly influential thirteenth-century encyclopaedias – and thereby into other forms of literature.

Following Jacques de Vitry's brickbats, camels were seen more and more – albeit never exclusively – as vehicles to convey aspects of sinful behaviour, most notably pride, anger and avarice. Petrus Berchorius (359b) compares the two types of camel, the Bactrian and the dromedary, respectively to visible and invisible pride; and the *Etymachia* equates the speedy dromedary to a proud man hurtling towards sin. That same fourteenth-century text also associates the camel with *ira* – because it makes clean water muddy and is implacably vengeful. Furthermore the beast represents avarice in a Palm Sunday sermon by the late thirteenth-century Franciscan Conradus Holtnicker (no. 99), and in a celebrated tapestry made for Regensburg Town Hall (see Harris 2007, 121).

No doubt the reasons why the camel attracted these precise associations are manifold; but it seems highly likely that they are based, however indirectly, on observation of real-life animals. To many observers, camels look contemptuously aloof, indeed arrogant; they are swift to anger; and their humps might well appear reminiscent of the zealously stored money bags of an avaricious person. In short, the more one knows about camels, the more miscast they seem as symbols of Christ.

VII

Slightly surprisingly perhaps, the camel does not appear in the *Physiologus*, possibly – after the Bible – the best-known book in early and medieval Christendom, and the source of numerous animal allegories that remained widespread and influential throughout the Middle Ages and, in some cases, beyond. Widespread and influential, but not unchanging. On the contrary, the *Physiologus* material underwent over the centuries a significant degree of modification, extension, contraction and adaptation. Such developments were not unique to the thirteenth century, but those adjustments to the *Physiologus* tradition which can be dated between roughly 1200 and roughly 1300 support our claim that these decades were of particular importance for the understanding of animals in Western culture.

Again it makes sense to take two contrasting examples. The first of these is the pelican, whose *Physiologus* association with Christ remained – in both the literal and figurative senses of the term – iconic over many centuries. The pelican, we are told, loves its young very deeply; but when the chicks grow and begin to strike their parents in the face, the parents become angry and strike back, killing their young. The parents feel great remorse and sorrow about this; and after three days the mother – in certain accounts it is the father – tears open her own side, and her blood drips down on to the dead young and revives them. In the same way, Christ's blood (and water) dripped down from the cross to revive and save us, his children (Gerhardt 11f.).

For many medieval (as well as modern) readers this ancient exemplum – however useful and indeed revered – raised a number of problems. Firstly, it is in no sense scientifically plausible – something that may not have mattered much in say, the first millennium AD, but which did in many thirteenth-century circles. Albertus Magnus, for example, dutifully recounts the familiar story before adding, more than a little sniffily, that "these matters are more read about in stories than proven by experience through natural science" (1642). Dante's teacher Brunetto Latini, in his *Livres du tresour* (c. 1260–6), tries to solve the problem more creatively, by claiming not that the parent pelicans' blood revives dead chicks, but rather that it keeps weak ones alive (I, 166). This attempt to combine scientific plausibility with moral theological interpretability seems to have satisfied no one, however; certainly I have seen no evidence to suggest that Brunetto's attempted solution won any followers.

A second problem for thirteenth-century users of the *Physiologus*'s pelican chapter was the motif of the angry parents killing their young. Unlike
the notion of parental self-sacrifice, this is rather difficult to integrate per-
suasively into a Christological interpretation. Hence we encounter, at least
from Thomas of Cantimpré onwards, a variant version according to which
the pelican young are killed by a snake which infiltrates its way into the
nest, much as Satan does into the Garden of Eden. Thomas introduces
this briefly, even tentatively:

> This bird loves its chicks, yet when they are importunate, it kills them and,
> having mourned them for three days, striking itself in the side with its own
> beak, she revives them with her own blood. It is said, however, that pelicans
> do this when they find that their young have been killed by a snake which
> has attacked them (V, 98).

The advantage of such an amendment, not least to a potential preacher,
is clear: if the snake (Devil) is the killer and the pelican (Christ) unequivocally the saviour, then the exemplum can be made to fit perfectly into the
conventional structure of the history of salvation. No doubt for that reason Thomas's alternative version proved popular in the later Middle Ages:
there are examples of it being used by John of San Gimignano (IV, 39), in
the *Etymachia* (160) and in two sermons by Johannes Gritsch (38.X,
43.U), as well as in numerous vernacular German texts catalogued by
Schmidtke (367–70). It is nevertheless instructive that, even in what is for
the most part a serious zoological work, an author like Thomas allows
himself the freedom to alter what is presented as natural historical fact in
order the better to accommodate a particular theological moralization.
Any modern notion that, in an exemplum intended for moral instruction,
the story or *proprietas* is primary and its interpretation only secondary flies
in the face of the practices of most medieval authors – even those affected
by, or indeed heavily involved in, an Animal Turn.

A final problem some thirteenth-century people seem to have had with
the *Physiologus*'s report of the pelican was its length. It is no more detailed
than many others from this source, even when all its details are subjected
to point-by-point allegory; but it is on the long side to be included in its
entirety in a sermon, and still more so in a lyric poem or fast-moving narrative. One suspects that considerations such as these are behind the trend,
increasingly observable as the Middle Ages progressed, to reduce
Physiologus and other such stories to their essential minimum – with the

result that allegories become, in effect, metaphors. This phenomenon is perhaps at its most effective in the celebrated thirteenth-century Corpus Christi hymn attributed to Thomas Aquinas, *Adoro te devote*. Its sixth section begins with the four vocatives "Pie Pellicane, Iesu Domine", which summarize the main purport of the *Physiologus* pelican chapter, before relating this, similarly economically, to the salvific power of the Eucharist: "Lord Jesus, good Pelican, make clean my filthiness and cleanse me with your blood, one drop of which can save the whole world from its sins". This is not the only example, though: Gerhardt (29f.) additionally refers to some vernacular texts where the pelican is reduced to a one-word metaphor for Christ: in Frauenlob's *Marienleich* Mary describes herself as "of the noble, worthy Pelican's blood"; a mystical treatise from the Rhineland urges the devoted soul to "enter into the sweet, wounded heart of the loving Pelican and see how he nourishes his children"; and, last but not least, Dante (*Paradiso*, XXV, 114f.) says of St John that "he is the one who once lay at our Pelican's breast".

VIII

In spite of these and numerous other changes during its story's long career, the pelican almost invariably retained its original association with Christ: it is telling that this identification is retained in all nineteen examples culled by Schmidtke from German-language literature between 1100 and 1500. In the case of the panther, however, the thirteenth century did not only witness tweaks to or transformations of the narrative part of the *Physiologus*, but also far-reaching changes to what the animal meant – a process for certain triggered by the Fourth Lateran Council's encouragement of animal exempla in the context of religious instruction, but one which at times took on an unpredictable and somewhat bizarre life of its own.

The *Physiologus*'s section on the panther is if anything still stranger than its account of the pelican. The panther has a beautiful, multicoloured coat and is by nature gentle and merciful. When it has eaten, it repairs to its cave and sleeps for three days. Upon waking, it roars, and from its mouth there emanates a wonderfully sweet odour. Hence, when the other animals hear and smell the panther, they hasten towards it – with the sole exception of the dragon, which cannot bear the panther's smell and in consequence remains in its own cave, where it lies as if dead. In the accompanying spiritual interpretation, the panther is standardly compared to

Christ, who after three days' "sleep" rose from the dead and saved mankind through the sweetness of his grace. The dragon, inevitably, is interpreted as the Devil, whose power was defeated by Christ's death and resurrection.

None of this behaviour, of course, has ever been observed by zoologists. Indeed, it is now universally agreed that the panther does not even constitute a species in its own right, but rather is just "an ordinary leopard carrying a recessive gene for melanism" (Morris 143). Albertus Magnus will not have known this; but all the same he was not impressed by what he read in the *Physiologus*. Hence he offers a highly naturalistic account of the *proprietas*, before asserting that it is, in fact, impossible:

> Due to the heat of its hunger, this animal, as do the other sharp-clawed animals, often overeats. At this time it takes itself to its cave and sleeps for a long time. When it awakens, a fine-smelling vapor emits from it and, according to Pliny, other animals follow this odor in herds. But… this is false, for no other animals apart from the human take either joy in or are saddened by smell (1531).

Nevertheless the *Physiologus*'s panther exemplum proved very popular in the Middle Ages and – unlike the story of the pelican – needed no new, evil character to be introduced before it could be conveniently used to illustrate Christ's victory over the Devil. The third aspect we noted earlier with regard to thirteenth-century treatments of the pelican does, however, apply also to the panther story – namely a tendency to abbreviate it radically, and in so doing to reduce it to one or two key elements. Several texts single out for attention the motif of the panther's enmity with the dragon. In two songs by the later thirteenth-century poet Konrad von Würzburg, for example, the panther's breath is presented as being not just disagreeable, but fatal to the dragon. In one of these we read that the panther is like Christ in that "through his smell, rich in many sweetnesses, he sees to it that the dragon lies dead before him, without resisting. In the same way the dragon of hell [the Devil] was defeated and felled" (*Kleinere Dichtungen* III, 1, 89–96). A second poem by Konrad, however, casts the panther as a munificent Christian, but the dragon as a sinfully avaricious person who cannot stand the holy smell of a generous person's praise, and hence flees (III, 26, 253–5).

This tendency to move away from meanings connected with the history of salvation and towards a more practical, moral theological orientation – a widespread tactic also in sermons – is seen later on in the work of the fourteenth-century mystic Heinrich Seuse (11f.), for whom the animal's breath is like the wisdom of holy scripture, which attracted him in his youth just as the panther "emits his sweet smell and draws the animals of the forest to him"; and in the *Etymachia,* where the panther's breath represents the attractive nature of humility, which draws people to it and indeed endows them with the other virtues (142f.). In marked contrast to the pelican, however, already by the second half of the thirteenth century we can find no shortage of examples of the panther's breath being interpreted in an unequivocally secular way. In the prologue to his *Conte du Wardecors,* Baudouin de Condé praises a rich man, who might justly be compared to "the sweet breath of a panther, whom the beasts seek and follow. Thus also minstrels follow the rich man wherever he goes… It is his sweet breath that brings comfort to minstrels" (I, 67–71, 75f.). Meanwhile "Der alte Meißner" locates the generosity of Margrave Albrecht of Brandenburg specifically in his "sweet voice", which has "the virtue of a panther", with the result that all indigent artists follow him (XVII, 11, 5); and "Der junge Meißner" attributes a panther-like "sweetness" to his patron Count Ludwig von Öttingen without limiting this quality to a particular characteristic or body part (A.II.1, 5–7).

Of particular interest, however, not least in that it raises questions that will recur in our next chapter, is a tendency amongst thirteenth- and early fourteenth-century poets to associate the panther with a human female – holy or otherwise. We can begin again with Konrad von Würzburg, and in particular his extensive Marian poem *Die goldene Schmiede* (602–5). Here the panther is compared to Mary, though Konrad's references to the Virgin's clothing and to the seasonal nature of the panther's exhalation bring overtones also of secular love poetry: "In Maytime all the wild animals run after the panther because of its sweet smell: in the same way many a soul hastens towards the odour of your clothes". These implications are drawn out clearly by Frauenlob (d. 1318) who, unusually, draws on the panther's potential negativity when he laments that his beloved draws him to her as the panther does the other beasts, who "follow him, because of his sweet smell, into bitter distress" (I, XIV, 1–4).

Frauenlob's slightly jaundiced simile is by no means the only example of a medieval poet casting a secularized, feminized panther in the role of his beloved. It is, however, unique in the sense that it is in German. Otherwise, this trend was the preserve of Romance poets. The reason for this is almost certainly grammatical gender: with very few exceptions (such as Baudouin de Condé in the passage cited above), French and Italian poets plainly regarded *panthère* and *pantera* as feminine nouns; but the commonest medieval German term, *pantier*, is not only neuter, but also explicitly combines the concept of an animal (*tier*) with the idea of "all", from Greek *pan*. Hence one might infer that using the animal to mean a woman will have seemed less obviously appropriate than it did to Romance authors.

The influential *Bestiaire d'amour* of Richard de Fournival (mid-thirteenth century) seems to have inaugurated this Franco-Italian tradition of interpreting the panther as the female beloved. Rather like Frauenlob, Richard's narrator reflects wistfully on his treatment at his lady's hands: she has caught him in a trap by virtue of his sense of smell; moreover "she still continually holds me at her mercy by smell, and I have abandoned my will in order to obey hers – just like the animals who, once they have smelled the panther's odour, can no longer distance themselves from her but, on the contrary, follow her unto death because of the sweet aroma that escapes from her" (145). In Sicilian and Tuscan lyrics of the Duecento, however, the panther's breath is an entirely good thing. Inghilfredi says of his mistress: "I shall never leave her; her wisdom makes me prudent, she holds me spellbound with her breath as the panther does the wild animals" (Jensen 94f.). Meanwhile Guido delle Colonne tells his inamorata that her mouth "exhales a perfume that is more fragrant than that of a wild animal called panther" (263); and Chiaro Davanzati offers a slightly different slant in opining that his beloved surpasses all other women "just like the panther with its perfume surpasses all other wild animals in gracefulness" (263).

By far the most extended and remarkable appearance of a panther in the role of lover is to be found in *Le Dit de la panthère* by Nicole de Margival, which since the work of Pierre-Yves Badel (especially 154–6) has generally been dated to the 1290s. Nicole's panther is a very rare example, outside the fable and beast epic traditions, of an animal who becomes an active, speaking character in a narrative (albeit one cast in the form of a dream); and she is unique in doing this solely on the basis of attributes assigned to her in the *Physiologus*.

Nicole's narrator states that he was in bed on the eve of the Feast of the Assumption (a Marian feast!) when he dreamed of being transported by a bird to a forest "full of many diverse beasts" including, above all, a *beste* standing at the entrance to a valley – a location, we are later told, that reflects its humility (628–36). This creature is of indescribable beauty, not least because its coat has in it something of the colour of every other animal; and all of these, save the dragon, want to approach it "for love of its sweet breath, which is sweet, good and healthy" – so much so, indeed, that it "can cure them of all their ills" (55–87). The narrator is mightily impressed by all this, but has to await the arrival of the god Amor for the animal's full meaning to be explained. It, or rather she, is called "une panthere" (451); her many colours reflect "the abundance of virtues that dwell in her", and her beautiful and curative breath signifies "her words, which are neither foolish nor mad, but wise, temperate and well governed by reason" (472–4, 499–502). These are able to heal such sins as pride and envy, which is why those who follow the example of the dragon, that is, "the toxically and harmfully envious" (523f.), flee from the panther as if for their lives.

The gradual transformation of the panther from an allegorical figure into the poet's beloved moves on a stage when the narrator, like so many other courtly lovers before him, at length meets her, but cannot think of anything to say (701–4). The metamorphosis becomes truly manifest only rather later, however, after the poet, spurred on by Amor and eloquently supported by various personified virtues, has visited the panther again and received from her the following response to his avowals:

> Since Pity, who is my mother, wants it, I will not refuse it; I want it also, and always will. And since Goodwill pleads for it, as well as Grace and her company, whom nobody should refuse, I do not know how to deny them; for I believe that they would not plead for anyone they did not know to be worthy: for this reason I do not dare refuse them. So you need not be sorrowful, because I grant myself and Mercy to your will in all goodness and all honour, without ill thought or dishonour (2156–71).

This rather repetitious declaration is, however, as far as the proposed interspecific relationship between the poet and his panther is allowed to get – for, a few lines later, the former wakes up, and his dream evaporates.

Nicole de Margival's strategy of fashioning a full-scale assimilation of the panther into his narrator's lady is a daring and in many ways fascinating literary experiment, which strays a good way from the *Physiologus* story on which it is nevertheless so obviously based. Nicole leaves us with no more than clues as to what precise broader cultural questions he might want the reader to reflect upon. Certainly it is a striking innovation on his part to equate the panther's breath with the words that she speaks – an anticipation, perhaps, of the association made by Dante, in *De vulgari eloquentia*, between his putative "illustrious language" and the panther – the "creature whose scent is left everywhere but which is nowhere to be seen" (I, 16). Otherwise it is worth nothing that Nicole's panther is used – albeit by a male author – as a means of constructing female aristocratic identity; and that the prospect, however fleeting and oneiric, of a *relation amoureuse* between a man and a panther points powerfully to the potential fluidity of the accepted boundaries between man and beast. These two aspects – the use of an animal to reflect and reinforce identity and an evident eagerness to challenge the orthodoxies of the "anthropological difference" – were both profoundly typical of the thirteenth-century Animal Turn, not least as it affected privileged members of the laity; and hence we shall encounter several further examples of both in the next chapter.

CHAPTER 4

Animals and Thirteenth-Century Chivalric Identity

Abstract This chapter assesses the many contributions made by animals to thirteenth-century attempts to construct or reconstruct chivalric identity. It discusses the keeping and giving of exotic animals by princes, to express their status, wealth or power; the use of animal symbols in heraldry and heraldic literature; the relationship between individual companion animals and their knights' identity and development (in *Iwein, Gauriel von Muntabel* and *Wigamur*); and, especially, the relationships between knights and their horses in a variety of epics and romances. This analysis shows that human and equine identities can become so intertwined and mutually dependent as in effect to merge – a phenomenon expressed particularly eloquently in the popular thirteenth-century figure of the courtly hippocentaur Chiron.

Keywords Aristocratic identity • Centaur • Chivalry • Horse

I

Hitherto we have dealt with an Animal Turn that primarily affected university, monastic and other clerical circles, that is, the world of learned Latinity. Not, of course, that this existed in isolation from the rest of European society: we have already discussed ways in which these developments influenced certain types of vernacular literature; the ultimate aim of

© The Author(s) 2020
N. Harris, *The Thirteenth-Century Animal Turn*,
https://doi.org/10.1007/978-3-030-50661-2_4

most nature exempla and some encyclopaedic writings was to communicate Christian truth to the laity via preaching; and one of the leading thirteenth-century ornithologists was, after all, a layman – albeit an Emperor writing in Latin.

Nevertheless it is time now to focus more firmly on the world of the secular aristocracy, specifically on the world of chivalry. According to Richard W. Kaeuper's persuasive chronology, the thirteenth century concludes the second phase of medieval chivalry (c. 1050–1300). During this phase knights – and their horses – continued to be active warriors; but the former also developed the "valorizing and inclusive ideology" (Kaeuper 85) that was courtliness; and their numbers in practice swelled, so that "by the later thirteenth century new social levels were incorporated within the elite of old and established ranks. Above the peasantry in France strode (with increasingly bold steps) gentlemen, knights not yet formally dubbed, those who had been dubbed, barons, counts, and dukes" (Kaeuper 119).

Alongside this latter development, and largely in contestation of it, the thirteenth century also witnessed what Maurice Keen (1984, 144) calls a "hardening of the rules governing admission into the order of knighthood, limiting the privilege expressly to those who could show knights in their ancestral lines". More so than previously, lineage became the crucial factor in establishing a person's nobility. By way of example, Keen points to an ordinance of Frederick II conveying this precise message, and also observes that the thirteenth-century version of the rule of the Knights Templar insists, unlike earlier versions, that no one should be admitted to the order unless he could show "that he is the true son of a true knight and a lady of gentle blood, and that he is descended on his father's side from a line of knights" (1984, 144).

Identity, status and lineage were, then, highly important but also fragile things, as Kaeuper's second phase of chivalry approached its conclusion. It is therefore almost inevitable that aristocratic self-representation and self-assertion should have become a conspicuous feature of thirteenth-century culture. Given the keen and widespread interest in animals we have alluded to elsewhere, and of course the *chevalier*'s (or *Ritter*'s) intrinsically close connection to at least one non-human species, it is also to be expected that animals would be involved in the construction of numerous forms of chivalric identity. And they were – be they real animals, depicted ones, literary ones, or fantastic ones. So in what follows we will consider in turn the role of exotic animals in menageries or similar contexts; of animals used as heraldic insignia or knights' companions; and, finally, of animals whose own identities implicitly merge with, or even are metamorphosed into, those of the knights with whom they come into contact. All of these phenomena, it can be argued, demonstrate that in the thirteenth century an Animal Turn was taking place in castles, battlefields and tournament grounds, as well as in the more rarefied air of churches, monasteries and universities.

II

The keeping of animals for purposes other than food or the performance of domestic and agricultural duties certainly predated the thirteenth century (quite apart from pre-medieval models, Henry I of England kept lions, camels and porcupines in his Royal Park at Woodstock from as early as 1104); but menageries are nevertheless particularly associated with the first two thirds of that century. Henry III famously established the Tower of London Menagerie in 1235, in response to Emperor Frederick II's gift of three leopards, and by the 1250s was accommodating there at least one polar bear and an African elephant – the latter, a gift from Louis IX of France, being the subject of two famously naturalistic drawings made by Matthew Paris in 1255.

The animal-keeper par excellence, however, was Frederick himself. He was not only a passionate hunter and naturalist, but also an inveterate collector of animal exotica, housed at various locations in Southern Italy and regularly accompanying him on his travels. An invaluable register for the winter of 1239–40, summarized and evaluated by Martina Giese (123–30), reveals that Frederick possessed an elephant, a giraffe (believed to be the first to appear in Europe), camels and dromedaries, horses and mules, bears, dogs, apes, lions, leopards (and/or panthers, and/or cheetahs), lynxes, a parrot, bearded owls – as well as, of course, innumerable birds of prey for use in falconry. This register also confirms that Frederick relied for the care and maintenance of this menagerie on an enormous roster of often highly specialized staff from all over Europe (there are references to 53 falconers alone).

So why did he do it? His Aristotelian interest in observing and studying the natural world – and in enabling others to do the same – must surely have been part of it. Clearly also Frederick was keen to engage in the giving and receiving of animal presents for political ends – an ancient practice which, however, had become much more frequent and elaborate, and had taken on a certain competitive edge, as a consequence of Westerners' progressively greater experience of the Orient. Giese (138–43) tells us, for example, that Frederick received Scandinavian gerfalcons from England, and for his part sent Henry III not just three big cats, but also a camel and some prize horses. Moreover his dealings with the Egyptian Sultan, before the latter's death in 1238, were still more extensive and complex: it was from the Sultan that Frederick received his elephant, giraffe and parrot, as well as horses and mules; and in return Frederick gave the Sultan a bear, a white peacock, an enormously expensive gerfalcon and – last but not least – his own horse.

The giving and receiving of such presents was, one can assume, a matter both of practical politics and of status-conscious self-representation. Similarly Frederick's passionate commitment to falconry must have been also – in part – a way of asserting his social superiority. As Robin S. Oggins (43) has put it, "falconry was an almost perfect example of conspicuous consumption: it was expensive, time-consuming, and useless, and in all three respects it served to set its practitioners apart as a class" – still more so, one might add, if practised on so lavish a scale as it was by Frederick. Above all, though, the very act of keeping, controlling and as such domi-nating some of the largest and most ferocious of beasts was in itself a potent assertion of Frederick's identity as the (would-be) secular head over all Christendom. A man who can control – and for that matter trans-port – a fauna consisting of such diverse and formidable animals as lions, elephants and camels must be fearless, able, powerful and enjoy a sphere of influence that far transcends the merely local. As Giese puts it (135), Frederick, through his menagerie, presents himself unequivocally as "a knower of the unknown, a ruler of the unruled, a tamer of the untame-able" – as, in a word, superior to all other princes, and as such uniquely qualified to rule over them and their subjects.

III

Such lavish assertions of chivalric and political identity were, of course, the exclusive preserve of the very rich: Oggins tells us (43) that, in thirteenth-century England, even a single falcon might cost as much as half of a knight's yearly income. Hence one suspects that the only "luxury" animals most knights could afford would be those painted as heraldic symbols on their shields. In itself, though, heraldry, "the systematic use of hereditary insignia on the shield of a knight or nobleman" (Keen 1984, 125) was a highly effective means of constructing and proclaiming a knight's chivalric identity – not least in a period like the thirteenth century, which attached great importance to birth and lineage.

Heraldry and heraldic animals were very much at their high-medieval zenith during the thirteenth century. They had developed out of the straightforward practical need to tell individual knights apart from each other – from the twelfth century onwards the bodies of combatants on horseback tended to be entirely encased in armour, and their heads and faces obscured by helmets. This problem extended beyond the battlefield, and became particularly acute, with the rise of the tournament in the second half of the twelfth century. Due to tournaments' particular focus on the achievements of individual knights and the often tightly packed nature of their mêlées, organizers, judges and participants needed a swift and reliable means of identifying which knights had been the ablest and most deserving of prizes.

By around 1200, the distinguishing devices displayed on knights' shields came to appear also on their helmets and tunics, on their horses' trappings, and – away from the field of combat – on their seals, tombs and effigies. They also ceased to be seen as relating exclusively to individuals, but rather as constituting hereditary symbols of chivalric families – and not always those of the very highest rank. In Maurice Keen's words (1984, 128), "heraldry,… from being originally the preserve of the greater aristocracy, came in time to be emblematic of the pride of birth, station and culture of the nobility in its broadest range". Like many other things in the thirteenth century, in other words, heraldic devices came to be badges of lineage, and as such instruments of both inclusion and exclusion.

Animals were depicted on shields from an early stage, particularly lions, eagles and panthers, and occasionally fish, falcons, ravens, stags, wolves and boars. From around 1200, however, the number of species involved increased considerably. An immensely useful list prepared by Heiko Hartmann (150f.) shows that the first attested heraldic use of a bear, a gryphon, a swan and an ibex occurred around 1200; of a steer in 1209; of a dog in 1210; of a crane in 1217; of a dolphin around 1220; of a peacock in 1223; of a hen in 1232; of a unicorn in around 1250; and, perhaps surprisingly bringing up the rear, of a horse in 1264. It is striking how many of these could be described as in some way exotic, or at least unknown in Western Europe – a consequence, no doubt, of the more detailed and nuanced perspectives on the animal kingdom that were themselves a feature of the first half of the thirteenth century; but also an indication that heraldic animals are in essence adstractions, "particularized examples of a general notion of [say] "the lion", a notion that recognizes the qualities associated with the animal without necessarily knowing the animal" (McCracken 76f.).

The specific relationship between these animal emblems and the individuals or families with whom they were associated is by no means always clear – we are, after all, dealing with exclusively pictorial symbols unaccompanied and unelucidated by written text. In many cases, no doubt, a merely generalized sense of an animal's ferocity and high status will have sufficed to convey the requisite message. Often also a particular animal will have been chosen because of a similarity between its name and that of the family it represents: it is surely significant, for example, that the first recorded heraldic uses of the swan, the dolphin and the elephant were by dynasties called respectively Schwangau, Dauphine and Helfenstein (Hartmann 150f.). That said, it is reasonable to surmise that, particularly as the thirteenth century progressed, some knowledge of the kind of clerical literature we covered in our last chapter will have penetrated far enough into the world of the lay aristocracy to influence choices of heraldic animals. So a nobleman choosing, say, the lion as his family's emblem could have been influenced to make this association by several factors. The lion's reputation as a ferocious predator; its status as king of the beasts; its frequent interpretation in Christian contexts as Christ or as various virtues; or its impressive, charismatic appearance, combining as it does the mane's ostentatious masculinity with the body's athletic yet intimidating elegance; some or all of these features may well have struck him as appropriate to the specific kind of aristocratic identity he wanted to construct.

For all their essential visuality, animal heraldic devices often constituted important motifs in thirteenth-century literature. Courtly authors no doubt used them in part because they were popular with contemporary audiences; but they were plainly also aware of their enormous literary potential: "Imaginary heraldic images allowed them to individualize figures, denote allies and relatives, give structure to potentially confusing mass scenes, especially descriptions of battle, create plot-shaping scenes of recognition and misrecognition, and add poetic life to events and dialogues" (Hartmann 168).

The possibilities opened up specifically by animal emblems were recognized and exploited in the early decades of the thirteenth century by Wolfram von Eschenbach, an author who knew more than most about the practicalities of contemporary chivalry. Several occur in his *Parzival* (c. 1205–10), where they "become synonyms of their bearers and themselves appear as participants in the action" (Hartmann 170, with examples). A particularly good instance of a heraldic emblem potentially operating on a variety of levels is the dove borne by the knights of the Grail Kingdom: on the basis of the bird's biblical associations it can be seen as reflecting their inspiration by the Holy Spirit and/or their role as what we would now call an international peacekeeping force; in the light of later clerical traditions the dove no doubt alludes to their essential chastity; and in the particular narrative context of *Parzival*, the emblem points to the key motif of a dove laying upon the Grail, every Good Friday, a Eucharistic host brought down from heaven.

If anything, however, Wolfram's use of heraldic animals is still more imaginative in his slightly later *Willehalm* (c. 1210–20), where they are combined with other kinds of animal image to create a veritable network of recurring motifs that illuminate aspects both of *Willehalm* and of *Parzival* (see Harris 2002, 215–19). A good example is the swan on the shield of the young heathen knight Josweiz, who is prominent in the second of the two battles the narrative describes. An attentive hearer or reader – of the kind that Wolfram plainly envisaged (see Volfing) – will have been reminded by this reference of an earlier use of the swan in relation to Galafré, another young heathen knight who fights in the first battle. He is described as "whiter than a swan" (27,1) – in complexion, one assumes from the context. Precisely how the reader is to interpret this recurring swan motif is left open: possible linking characteristics are the two knights' youth, their virtue, or their role as courtly lovers. Moreover the device suggests a link between the narrative's two battles, which are more closely linked thematically than immediately meets the eye. The swan is, in effect, one of a number of devices Wolfram uses in order subtly to imply this interconnectedness from an early stage. Moreover a reader of *Willehalm* with any knowledge of *Parzival* is likely to forge a mental link between Galafré, Josweiz and the swan knight (and Parzival's son) Loherangrin, who is introduced at the very end of that romance and who possesses a comparable mixture of youth, knightly excellence and noble birth. That reader might then also recall that, in *Parzival* 257,13, the narrator has described the skin of the beautiful Lady Jeschute using exactly the same words ("whiter than a swan") as those applied, in *Willehalm*, to the handsome knight Galafré. There are plenty of places within the narrative universe inhabited by these two great works where Wolfram more obviously challenges conventional gender divisions and stereotypes; but he also does so effectively here; and the use of such precise verbal parallels plays an important part in helping him to establish a rich and coherent narrative texture.

I cannot think of any other medieval author whose heraldic animals are deployed with such refinement or attention to detail; but their popularity amongst courtly and, increasingly, urban audiences is attested by their frequent appearances elsewhere. For example Wirnt von Grafenberg's *Wigalois* (1210s) considerably expands the repertoire of literary heraldic symbols, including references to eagles, a swan, a stag, leopards, dragons, a golden lion, gryphon claws and an elephant. It is hard, however, to discern in Wirnt's use of these any clear purpose beyond injecting some local colour and facilitating identification; and this judgement could be extended to cover most heraldic devices described in literature from the two generations following Wolfram. An exception perhaps is the enormous romance *Diu Crône* (after 1225) by Heinrich von dem Türlin. Heinrich's narrator sometimes tells us why a particular knight has chosen a particular heraldic device: Galaes, for example, has the fearsome claw of a black bear on his shield, "from which one should learn that he is fiercer than a bear" (9812–17). Moreover Heinrich is unusual in spotting the comic potential of knights going into battle with animals on their shields. Gasoein de Dragoz's shield, and also surcoat, feature a ferocious-looking golden lion; and, in the heat of battle (10544–57), this lion comes to life, roars prodigiously and sticks its tongue out, before returning swiftly to the two-dimensional painted existence from whence it came.

One further work from thirteenth-century Germany (specifically 1257) must, however, be mentioned in any account of the literary career of animal insignia, namely *Das Turnier von Nantes* by Konrad von Würzburg (see *Kleinere Dichtungen*, II). This work of 1156 lines is a remarkable example of what came to be known as "heraldic poetry", in that it consists in large measure of descriptions of the armour, equipment and heraldic emblems worn by the participants at a fictitious tournament in Nantes. The tournament's climactic second stage involves fighting between a group of French and Spanish knights under the leadership of the unnamed King of France, and a collection of German knights led by one Richard, King of England. The latter are victorious, and the star of the show is in every respect King Richard. Over twenty animal emblems are mentioned in all. These are not used with any great literary creativity, but they are carefully and systematically described (always for example including a reference to their colour), in ways that perhaps reflect, and certainly anticipate, the mainly fourteenth-century catalogues of emblems known as armorials, or heraldic rolls. In Konrad's poem, as well as in the roughly contemporaneous early rolls such as that of Matthew Paris (c. 1244, reproduced by H. Hartmann 167, and Scheibelreiter 187), one can already see the same desire, albeit on a much smaller scale, to draw together, organize and systematize information that we observed earlier in the cases of Aristotle, the thirteenth-century encylopaedists and the collectors of nature exempla.

A further important feature of *Das Turnier von Nantes* is that, like the *Pavo* of Alexander von Roes, it uses animals – albeit here exclusively heraldic ones – in the service of political point-making. Many of Konrad's attributions of animals to knights unequivocally reflect contemporary realities: the Margrave of Brandenburg really did bear a red eagle (cf. Konrad 424–39), the Landgrave of Thuringia a red and white lion (474–85), the Duke of Lorraine three white eagles (608–19), and the Margrave of Meissen a black lion (1002–5). Above all, the English king sports the three leopards (297–303) which – even though they have tended to be described as lions – have been part of the heraldic crest of the English crown since the reign of Richard the Lionheart, and which doubtless also inspired the feline nature of Frederick II's gift to Henry III in 1235. All of these references to the "real world" self-evidently lend Konrad's poem an air of truthfulness and authority, and they also provide readers with an accessible key with which to decode his political message – that, as in the tournament, the Low German princes should rally behind a heroic and virtuous English Richard. In 1257 that can only have meant King John's second son Richard of Cornwall, who was crowned King of the Romans in Aachen in the May of that year.

IV

We have already seen Heinrich von dem Türlin asking himself the question, tongue firmly in cheek, as to what might happen if an animal on a knight's shield suddenly came to life. One could argue that this question is responded to more seriously, and certainly at far greater length, in a series of thirteenth-century romances in which knights are accompanied by, closely associated with, defined by, or indeed (re-) named after animals who also had extensive heraldic careers. Probably the two best examples of this are the famous lion in Chrétien de Troyes's *Yvain* (c. 1180) and Hartmann von Aue's *Iwein* (c. 1205), and the much less celebrated eagle of the anonymous *Wigamur* (mid-thirteenth century). In both of these works – as well as, less concertedly perhaps, in the case of the goat in Konrad von Stoffeln's *Gauriel von Muntabel* (second half of thirteenth century) – we can see the animal in question adding much to an author's exploration of the eponymous hero's identity.

For reasons of chronology, and also because it enables us to cover three authors working within a specifically German tradition, we will focus on Hartmann's lion rather than Chrétien's. When Iwein first encounters it, at almost exactly the central point of the romance's action, his fortunes are at a very low ebb. He has committed the serious chivalric sin of staying away from his wife Laudine for considerably longer than he has promised; and for this he has been roundly denounced by her lady-in-waiting Lûnete. This in turn has plunged him into an existential crisis, in which he has traumatically lost both his self-respect and his sanity – as well, of course, as his knightly reputation. Thanks to God (3261), working through the agency of a page and then a lady-in-waiting, Iwein begins tentatively to recover his health and to re-enter the civilized pale, by successfully defending the lands and person of the Lady of Narison from the rapacious attacks of Count Aliers. Almost immediately thereafter, he encounters a lion fighting – and nearly being killed by – a dragon (or dragon-like snake). After some hesitation, caused not least by the fear that, if he were to save the lion, the latter might in turn attack him, Iwein intervenes successfully on the side of the "noble beast" (3849), and kills the dragon. Thereafter the lion becomes Iwein's constant companion, grateful, loyal and helpful both in combat and in more peaceful times.

Inseparable as they are, however, it always remains abundantly clear that the lion is an animal, and as such does not share in Iwein's humanity. Immediately after it has been rescued, for example, the lion "stretched out at his feet and by gesture and voice gave him a wordless greeting. It stopped its raging and showed its affection as best it knew how and as well as a beast could" (trans. Tobin et al., 3869–76); and soon afterwards it "scented prey" (3885), an instinct that it can express to Iwein only by "stopping, looking at him, and pointing with its muzzle" (at a nearby deer, 3890–3). Thereupon the lion kills the deer, Iwein cooks it, and together they eat it – in itself an eloquent statement both of their friendly companionship and of the fact that such companionship does not fundamentally challenge conventional animal/human boundaries.

That said, many aspects of the lion's instincts and behaviour do prove – allowing for the limitations and specificities of its animal nature – to be strikingly reminiscent of a courtly knight. It is the epitome of loyalty, accompanying Iwein everywhere, to the extent that the latter comes to be described exclusively as "the knight with the lion". It also seeks to protect Iwein, "staying awake, pacing watchfully around him and his horse. At all times, both then and later, it had the noble intention of guarding him" (3914–16; also 5226) – one suspects there are overtones here of the *Physiologus* conceit that the lion always sleeps with its eyes open, just as Christ never ceases to watch over his children. Iwein's lion is also movingly self-sacrificial, both in seeking to kill itself when it believes its master dead (3940–9), and in courageously intervening on three occasions to support, indeed save Iwein in combat (5050–74, 5375–422, 6737–98). All of these interventions, however, are made in the interests of justice and chivalry, when Iwein is unfairly mismatched against ignoble opponents – successively the giant Harpîn, three human antagonists and two more giants. When he eventually fights the comparably chivalrous Gawain, by contrast, the lion is not there.

Still more so than that of a static heraldic lion on a shield – which, incidentally, Iwein's lion briefly becomes, when its companion carries it away wounded following the fight with the three men (5564–73) – the precise meaning of this dynamic lion figure is hard to pin down. It has been variously interpreted, for example, as Christ, as an emblem of power and/or justice, as an astrological symbol, or indeed as an idealized aristocratic wife. All of these attempts at an allegorical definition have some validity – particularly, perhaps, the notion of the lion as Christ (bearing in mind its readiness to die and the motif of its keeping watch). One wonders, however, whether such arguably limiting interpretations do full justice either to the complexity of Hartmann's lion image or to the precise contours of his narrative structure. The lion is one of the romance's major figures, but it first appears as late as line 3828 (of a work lasting 8166 lines), and its work is essentially done with the final victory over the two giants in line 6798. Thereafter its role is purely secondary, and indeed passive: it becomes simply a means by which Iwein is identified by those, such as Gawain (7762) and Lûnete (7950), from whom he has long been estranged. Moreover it is not mentioned at all in the climactic final scene in which Iwein is reconciled with Laudine and can again be called by his own name (7974, 8074, 8097, 8122), rather than being referred to only as "the Knight with the Lion".

One is forced to conclude, then, that the lion plays an important role in constructing Iwein's chivalric identity only in a particular, if substantial, part of the romance. Seemingly that identity does not require the presence

of a lion at the very beginning, when Iwein is a respected, fully function-
ing, albeit youthful knight; nor is the lion needed at the end, when the
hero is rehabilitated and accepted back into society. Nor, for that matter,
does it accompany Iwein in the darkest depths of his trauma, *before* his
decisions to help both the Lady of Narison and the lion itself. Immediately
thereafter, though, as soon as Iwein has signalled that his instincts, however weak and unstable, remain those of an authentic knight, the lion
begins as it were vicariously to supply him with those values and facets of
his own personality from which he is temporarily estranged – an awareness
of God, along with a sense of justice, loyalty, service, moderation and self-
restraint. All these are actually intrinsic parts of Iwein's, and any true
knight's, identity; to varying degrees, however, he has lost them, and for a
time he needs the lion's presence and help to re-access them and reinte-
grate them into his own personality. This process proves to be a complex
but eventually successful one, such that by the end of the romance Iwein
has regained a healthily independent – and by now more mature – personality, as well as a properly integrated social role. Hence at length, its job
done, the lion can recede from the reader's consciousness, as well as, to an
extent at least, from Iwein's own.

A further example of an animal figure being used in the construction,
or reconstruction, of a knight's identity at a certain key point in his career
can be found in *Gauriel von Muntabel*. The hero of Konrad von Stoffeln's
romance is in an amorous but secret relationship with a fairy queen, who
has imposed on him a taboo preventing him from mentioning their intimacy to anyone. During a frankly flirtatious conversation with another
lady (118–43), however, Gauriel does at least hint that he already has a
beloved, with the result that the latter soon afterwards curses him: "Beauty
will now be taken wholly away from you, and you will present an ugly
countenance, so that men and women will fear you. That will be my punishment for you. But you will keep physical strength, reason and your
manly character" (258–63). The fairy queen is as good as her word:
Gauriel suffers a seriously disfiguring six-month illness, after which he sets
out to try and earn his lady's pardon. As he does so, we are told out of the
blue – this is not always the most coherently motivated of stories – that
Gauriel has raised

> a large and strong goat, which never minded traveling near and far with its
> master and helping him out of many a tight spot in which he had seemingly
> been marked for death already. On his armor he bore a goat made of gold,

likewise on his shield. On account of this his name changed. He was called in all lands 'the Knight with the Goat' (trans. Christoph, 311–24).

So Gauriel goes on his way accompanied both by heraldic goats and by a "real" one, and proceeds to begin his knightly rehabilitation with a string of victories against minor Arthurian luminaries. The goat remains passive during these, but is called into action to participate in a bizarre combat in which it partners Gauriel in a four-way combat against Iwein and his lion (1820–2033). This results in the death of both beasts: the lion is killed by the goat, and the goat by Iwein.

Sabine Obermaier (2004, 135f.) is for certain right to read the goat's killing of Iwein's lion as a poetological comment on Konrad's part, through which he hints at a need to move beyond the by his time "classical" structures of German Arthurian romance as established by Hartmann. The fact that the goat also dies, however, surely indicates that the episode additionally draws a line under a particular stage of Gauriel's development. Rather than contrasting with and supplementing the deficiencies of its master, as Iwein's lion does, Gauriel's goat comes across as a relatively straightforward symbolic representation of the knight himself at this still relatively early stage of his career. The goat is, in the eyes of most people, an ugly animal, and since time immemorial it has been associated with lechery – or, in medieval terms, the sin of *luxuria*. At the same time, Gauriel's goat is described as fighting with enormous strength and courage, both of which qualities it palpably shares with its chivalric companion. Its physical demise does not spell the end of Gauriel's goat-likeness: he retains the goats on his shield and armour, and is still sometimes known as "the Knight with the Goat". Nevertheless the goat's death certainly implies that Gauriel's *luxuria* and its concomitant ugliness will before long be things of the past. And so it proves: as early as line 2517 he sets off – appropriately enough in the company of Erec – to return to his lady's land and seek her pardon; shortly after his arrival there she lifts the spell and he is restored to being "the most handsome man anyone had ever known of" (3119f.); and soon after that the reconciled couple are married (3245–7). Gauriel is free to continue his adventures without the ugliness and sinfulness his deceased goat has represented.

The eagle of *Wigamur,* meanwhile, participates in the construction (and in this instance revelation) of its knight's identity in a strikingly different way. Wigamur, the son of the king of Lendrie, is abducted when a small child by the mermaid Lespia and brought up under the sea first by her and then by a courtly centaur-like sea monster. When the latter eventually sends him out on to dry land, armed with a bow and arrow, Wigamur is keen to gain honour, but also to discover his origins and lineage. He embarks upon a series of adventures which reveal both his immense courage and his naive inexperience, before receiving a somewhat belated chivalric education at the hands of his uncle Ittra – who claims also to have been instrumental in the upbringing of King Arthur (1397–1401). Upon leaving Ittra, Wigamur sees an eagle feeding its young, one of which is promptly stolen and devoured by a vulture (1455–65). This gives rise to a ferocious battle between the two adult birds which, overcome as it is by the vulture's foul stench, the eagle seems about to lose (1480f.). Revealing a greater degree of clear-sighted mental acuity than did Iwein when faced with the lion, Wigamur immediately acts to save the eagle, killing the vulture with his bow and arrow. Having attended to its surviving young, the eagle reacts initially to Wigamur's intervention much as the lion did to Iwein's: sitting alongside the knight, it makes species-appropriate gestures that suggest gratitude (1489–93). As soon as Wigamur departs, however, the eagle immediately begins to fly alongside him; and, at least as far as the reader is made aware, it continues to do this for the remainder of the knight's career: "When Wigamur rode away, the eagle flew alongside him always, and kept on the same level as him wherever he rode or walked. Thus it never left him; it always flew to wherever it could find Wigamur's horse. All night the eagle kept watch, so that nothing might harm him" (1509–17).

The eagle behaves, then, like a curious cross between a guardian angel and an airborne heraldic emblem. Otherwise it does nothing. Certainly it has no need to intervene, as Iwein's lion does, either to save its companion or to make good any of his deficiencies. On the other hand, its role in Wigamur's life comes to no discernible end: at the triumphant apex of his career he simply changes from being "the Knight with the Eagle" to being "the King with the Eagle". Plainly, then, the eagle represents something that simply *is*, and has no need to change or to prove itself. In the light of this Obermaier (131) is surely correct to regard it as a symbol of Wigamur's (to him long unknown) regal origins, and hence also as a visible reminder of the importance of the genealogical principle in general. Given its universally accepted status as king of the birds, as well as its role as an imperial emblem, the eagle is ideally suited to conveying such a meaning; and the romance itself contains several motifs that tend to reinforce it. Immediately upon hearing of the approach of Wigamur's eagle, for example, Queen

Isopi becomes fully convinced that King Arthur will keep the promise he has made to her and that all will turn out well (3097–100). Royalty is at hand; and royalty can be relied upon.

Of these three companion animals, then, the lion and the goat on the one hand and the eagle on the other are employed to promote two contrasting conceptions of nobility and aristocratic identity. *Iwein* and *Gauriel* elaborate the idea forged in the creative melting pot that was the courtly world of the twelfth century, that one could become truly noble, and at the same time legitimize one's political or social ambitions, by dint of one's own behaviour and achievements – a process which tended to be lengthy and far from straightforward, but which was by no means restricted to those at the very top of the social ladder. The ideology of *Wigamur*, by contrast, with its strong emphasis on birth and lineage, is characteristic of the rearguard action undertaken in the thirteenth century by many members of the established aristocracy who sought to reassert the notion that nobility is a matter of birth – that a true knight can indeed only be "the true son of a true knight and a lady of gentle blood". These are, of course, ultimately irreconcilable ways of looking at a question which, by its very nature, can never be finally settled; but it is strikingly significant that, in the context of the thirteenth century, such a question is explored not least via the concerted and protracted use of animals.

V

Effective though lions, eagles and goats might be in enabling authors to investigate questions of personal development and chivalric identity, their importance to thirteenth-century knights pales into insignificance beside that of the horse. Knights simply could not do without their horses. They were, first of all, ubiquitous markers of a knight's status: Jordanus Rufus, in his thirteenth-century manual of equine training *La Marechaucie des chevaux*, reminds us that "no animal is more noble than the horse, since it is by horses that princes, magnates and knights are separated from lesser people and because a lord cannot fittingly be seen among private citizens except through the mediation of a horse" (quoted by Cohen, 46). Still more than that, though, horses formed an intrinsic part of the everyday life of the knight, essential for transport, for fighting, and hence for the attainment of all kinds of honour. In the end indeed one could argue, with Jeffrey J. Cohen (61), that "knightly identity depends more on animal bodies than upon mere heterosexual desire or quotidian social structures like family".

It does not take an enormous imaginative leap for us to understand that knights were often close to their horses in affective terms; Cohen's formulation above, however, reminds us that their intimacy was also by definition physical, and as such capable of including from time to time homoerotic dimensions (both knights and their destriers were by definition male). At the same time the physical closeness of horse and rider was increasingly facilitated, and rendered effective, by inanimate objects. A good example of this is the introduction of the stirrup into Europe in the seventh or eighth century, a device leading to "an enlargement of somatic possibility" which enabled the horse "to become more responsive to its rider within an augmented tactile syntax between equine and human flesh" (Cohen 49). Another such invention is the so-called "shock charge" in the eleventh century, which, when employed in battle, has the effect of gathering together "horse, rider and lance... into what has been called a 'human projectile'" (Keen 1984, 24). The "horse – knight – stirrup – spear assemblage" (50) of the shock charge is an example of the permeability of species boundaries on which the whole concept of chivalry arguably depends, and also of a "chivalric circuit", in which "the horse, its rider, the bridle and saddle and armor together form... a network of meaning that decomposes human bodies and intercuts them with the inanimate, the inhuman" (Cohen 76).

Not least in the thirteenth century, the relationships and borderlines between these various elements of the "chivalric circuit", and especially between its human and equine parts, were often explored in imaginative literature. For a relatively early example we can turn again to Wolfram von Eschenbach's *Willehalm*, an adaptation of the sixth instalment of the French William of Orange cycle, *La Bataille d'Aliscans* (probably 1180s). *Willehalm* describes an especially close relationship between the eponymous hero and his horse Puzzât. In the first two of the epic's nine books this pair form an effective chivalric circuit that also includes Willehalm's sword Schoyûse – which is in fact mentioned rather more often than Puzzât. Unlike the latter, however, Schoyûse is never engaged by its owner in conversation. At the very beginning of Book II, when he is on a mountain top looking down over the vast Saracen army, Willehalm turns to Puzzât and addresses him in the following terms:

Alas, ... Puzzât, if only you could advise me which way to turn! How I could use your strength if we were now fit and free from wounds! Then, if the heathens took it into their heads to chase me, some of their kinsmen would have cause to regret it. Now both of us are lame, and I am bereft of joy. You can be sure of one thing: if we get back to Orange, provided that the heathens have not taken it away from me, I shall make you happy with vetch and oats, peas, barley and sweet hay. Now I have no solace but you. May your speed bring me consolation (58,21–59,8; trans. Gibbs/Johnson 43).

This remarkable speech, part lament and part promise of reward, contains plenty of evidence that Willehalm regards his horse with affection and respect: he readily identifies with the mount's physical weakness, using the first person plural pronoun to indicate that he sees the two of them as a team or unit; he is plainly concerned for Puzzât's welfare as well as for his own; and he wishes to see him happy. At the same time, it is clear that Willehalm thinks of Puzzât unequivocally as an animal, with an animal's limitations: he knows that his horse cannot, in fact, advise him; and he implies that the only way he can help his rider to achieve solace and consolation is by running swiftly.

This impression that Puzzât, whilst highly intelligent and of great value to Willehalm, nevertheless has no real share in the latter's humanity is confirmed in subsequent scenes. Immediately after having heard the above speech and having been rubbed down by Willehalm, we are told that "Puzzât's weariness began to leave him: he snuffled and snorted, and thanks to Willehalm's care he recovered from the tremendous weakness which had come over him" (59,6–10; Gibbs/Johnson 43). He responds, then, both to his knight's pep talk and to his solicitous practical gesture, but does so in a quintessentially equine fashion. Later indeed, having been seriously wounded in combat, Puzzât proceeds to behave in ways that are almost more canine than either equine or human. Having killed – indeed decapitated – Arofel, the brother of the principal heathen potentate Terramêr, Willehalm takes his armour as a disguise, and his horse Volatîn as a replacement for Puzzât. This does not mean that he in any way abandons the latter; but he does attract rather more continuing attention from Puzzât than he has bargained for: "Puzzât was badly wounded, and so Willehalm at once took off his harness, so that he might find food for himself, but the horse followed him away from there, and, wherever his master rode, took the same path after him" (82,9–14; Gibbs/Johnson 54). This is of course touchingly loyal behaviour, but it swiftly becomes problematic for Willehalm because it results – even though he is wearing Arofel's armour – in his being recognized and challenged by enemy forces. Hence, when Puzzât is killed soon afterwards, Willehalm certainly laments his loss (88,23; Gibbs/Johnson 56); but the reader might well reflect that, in hard-headedly realistic terms, this is not an unmitigated disaster.

In the light of much of the evidence we have considered so far, one might expect than an early thirteenth-century treatment of a knight's relationship to his horse would tend to play down the anthropological difference between them rather more than a source text from a generation earlier. If this is indeed a trend, however, *Willehalm* and the *Aliscans* buck it. The French poet seems much more interested than does Wolfram in exploring the idea that his horse, Baucent, might possess human traits. Not only does his hero Guillelme talk to Baucent more often than Willehalm does to Puzzât, and at times almost as if he were a human vassal, rather than a horse; but the narrator of the *Aliscans* says of Baucent's response to the speech quoted above that he "understands him [Willehalm] like a wise man" (ed. Holtus 602f.). Passages like this, however, do not mean that Wolfram is less progressive, or indeed less interested in reflecting about animals, than his French predecessor. Rather it is telling that, throughout his adaptation, Wolfram systematically removes all motifs that tend towards humanizing animals and dehumanizing the heathen army. This reflects a clear decision to prioritize a different – and certainly no less forward-looking – agenda, namely that of establishing the essential equality before God of all human beings, Christian and non-Christian alike. In that revised context Puzzât has to remain an unequivocally non-human animal, albeit one to which Willehalm is both practically and sentimentally attached.

By contrast, other adaptors of French sources did choose to emphasize and expand any hints they found of a particular closeness or integration in the relationship between hero and horse. This is true, for example, of the Middle English *Bevis of Hampton* (c. 1300), which makes considerably more of such motifs than its predecessor, the Anglo-Norman *Boeve de Haumtone* (1190s). In particular, Susan Crane (154–7) has shown that the relationship between the English Bevis and his "good and loyal steed" Arondel is in many ways the most stable, faithful and productive in the romance – qualitatively superior, in other words, to many relationships between humans. The latter are frequently characterized by treachery and betrayal, whereas the watchwords for the bond between Bevis and Arondel are "love" and "loyalty".

Unusually for romance, the English author also offers implicit parallels between Bevis's relationship with Arondel and that involving him and his female beloved, Josian. This is done partly through motifs which link the career of the lady with that of the horse (see Crane 156), and partly through invitations to juxtapose the two figures' attitudes and behaviour. At one point Bevis himself compares Josian's loyalty negatively to that of his horse (2033–5); and a good deal is revealed about the pair when they simultaneously catch sight of Bevis on his return disguised as a pilgrim: "Bevis went up to the horse; then the horse saw and knew him. He stood completely still until Bevis caught hold of the stirrup. Bevis threw himself into the saddle; thereby the maiden well knew him" (2175–80; Crane 156). The order of events here maybe suggests that Arondel remains Bevis's "first love"; but it is also noticeable that the horse recognizes his master before he has "reasserted his position as a mounted knight" (Crane 156), and the lady only afterwards. Maybe, then, there is something innate, instinctive or visceral about the bond between knight and horse that is lacking from many, perhaps even all, relationships between humans. Certainly Crane (156) is right to suggest that,

> throughout the romance, the horse resembles Bevis more closely than he resembles Josian: like Bevis he is emphatically masculine, courageous in combat, and tireless on adventure. The bodily and temperamental likenesses linking Arondel and Bevis give rise to the loyalty, understanding, and love between them.

Relationships such as that between Bevis and Arondel are plainly profound and mutually symbiotic; certain authors, however, go further even than this by suggesting that knight and horse can share what we would now call a *Schicksalsgemeinschaft*, a relationship between two very similar beings who share a common destiny – in which the horse becomes, to all intents and purposes, the knight's alter ego. Using Philippe Descola's theory of the fundamental permeability and negotiability of the concepts of nature and culture, Lieselotte E. Saurma-Jeltsch (2017) has argued that just such a connection exists in medieval accounts of the Alexander legend between the eponymous hero and his horse Bucephalus. She supports this argument with reference mainly to the *Roman d'Alexandre* by Alexandre de Paris (c. 1180) and to visual sources; but the case can be further strengthened by giving a comparable emphasis to two thirteenth-century

German versions, those by Rudolf von Ems (c. 1250) and Ulrich von Etzenbach (1280s).

Alexander and Bucephalus have in common the fact that they are both hybrid creatures. Some of the more animalistic elements in Alexander's make-up stressed by earlier legends – such as his possession of ram's horns – are not found in the medieval versions; but Rudolf von Ems states, for example, that he has one yellow and one black eye (deriving respectively from an eagle and a dragon), the hair of a lion, and animal-like sharp, round teeth (1310–17). Bucephalus, meanwhile, is descended from a horse and a gryphon, has the head of an auerochs and a horn on his forehead (2103–16). Moreover he is much stronger than a lion or an elephant (2118f.), and is an enthusiastic man-eater: he eats people "as if they were hay" (2117), and is kept in irons – like a knight's armour? – until such times as his services are required for the purpose of devouring malefactors. The particular closeness of the relationship between Alexander and Bucephalus seems moreover to have been predetermined by the fact that they were born on the same day (Alexandre de Paris I, 424–9); and it is sealed by Bucephalus's instinctive appreciation of their bond when the two of them first meet. Hitherto wild and untameable, he immediately humbles himself, kneels down and permits Alexander to ride him.

Thereafter Rudolf von Ems points out several times that, in battle, Bucephalus fights much as a knight does: he is clad in irons (7386), and can kill human opponents independently of Alexander (2493–5); in short, "he fights like a man" (12588f.). This notion of Bucephalus as Alexander's "other self" is lent further substance by the fact that, when the latter fights a battle without Bucephalus, he loses; and the two remain firmly and seemingly intrinsically linked even after the horse's death. Alexander then becomes poignantly aware that "now he has lost his alter ego, his own death is near" (Saurma-Jeltsch 38); and Ulrich von Etzenbach in particular brings out the extent to which this bereavement is, for Alexander, a cataclysmic event: he is determined to avenge Bucephalus (20044–56), laments his death bitterly (21884–6), and returns to the battlefield on which he has died to give him a proper burial in silk cloths, and erect a gravestone commemorating "the deeds that he had done on the horse" (23529–55). Ultimately, then, for all this singling out of Bucephalus for particular celebration, Alexander's focus is on their role as a militarily outstanding interconnected dyad.

VI

In purely physical terms at least, the most organically connected human/horse dyad is of course the centaur (more accurately, the hippocentaur), with its human head and torso, and equine lower body and legs. Many have believed, after all, that the very idea of a centaur resulted from a misapprehension that nomads and their equine mounts constituted a single animal entity. Hence it is not surprising that, as thirteenth-century writers sought to explore the reality or otherwise of the human/animal divide, they should have turned their attention to centaurs – rather as slightly earlier French authors had done with the werewolf, a figure used most fruitfully in works such as the *Lai de Bisclavret* of Marie de France (see Crane 54–68) and the anonymous *Guillaume de Palerne* (McCracken 78–85), but whose medieval literary career seems to have been relatively circumscribed in time and place. By the time of Dante, certainly, centaurs seem to have been the most obvious beings by which to represent an amalgam of human and animalistic identity: significantly, he was the first author to interpret Virgil's Cacus (described in the *Aeneid* (VIII, 190–267) both as a *semihomo* and as a *semiferus*) as a centaur – albeit one whose rage and "crooked deeds" are symbolized by the presence of a dragon riding on his back (*Inferno*, XXV, 17–35).

In earlier medieval literature, the reception of centaurs was hindered to some degree by the *Physiologus* and bestiary traditions having associated, and sometimes confused, centaurs with sirens. Almost invariably treated in the same chapter, these two beings, with their dual natures, were commonly used as images of hypocrisy. It is once again to the post-Aristotelian encyclopaedic tradition – in this case specifically to Thomas of Cantimpré – that we owe some clarity in the matter. He simply but definitively assigns sirens to his Book VI, on sea monsters, and centaurs – whether they involve horses, asses or steers – to Book IV, on quadrupeds.

Aided no doubt by such greater transparency, but motivated also by the age's ongoing fascination with classical antiquity, we find no shortage of centaurs in thirteenth-century literature. Some of these reflect their genus's traditional association with unbridled ferocity. This is especially the case when the identity of a hippocentaur is combined with other elements to create a monstrous "strange creature". A good example of this is Marrien in Wirnt von Grafenberg's *Wigalois*, whose human upper body and equine lower body are accompanied by a dog's head, long teeth, a wide mouth and deep-set, fiery eyes; whose sex is impossible to discern;

whose body is covered in scales harder than stone (6931–47); and whose attitude towards the hero is violently hostile.

Outnumbering such "bad" centaurs, and of greater interest to our current context, are presentations of the "good" centaur Chiron, the tutor of Achilles, in whom there was a marked revival of interest in thirteenth-century Germany. In the *Achilleid* of Publius Papinius Statius (first century AD) Chiron is already a highly positive influence on the young Achilles; but in *Der Trojanische Krieg* by Konrad von Würzburg (c. 1282–7) he becomes both a superhumanly efficient fighter and a highly accomplished courtly gentleman – a figure, then, who integrates and also surpasses both the human and animal elements of the "chivalric circuit" and hence has much to tell us about the nature and possibilities of chivalric identity.

Konrad introduces Chiron (whom he calls Schyron) in some detail:

> He was called Schyron and presented a strange picture, for his wild figure was of a dual nature. The top part was formed like a man, and attached to it was his lower half, which was like a horse's. This accomplished man was stronger than all animals. Even the gryphon and the fell lion trembled at his force. He could master dragons and snakes. His battle frenzy drove away the frightened animals. He could wield a sword and shield better than any man. He was good for any ruse of battle. He was the best archer the world has ever seen. His entire skin was as hard as horn. No bird was so swift that it could fly more quickly than he could run (5850–75).

So fascinated is Konrad's narrator by this extraordinary, if exotic, paragon that he offers a second, and still more detailed description only a little afterwards (5918–85), in which the thoroughgoing extent of his dual, animal and human nature becomes still clearer. Chiron has long eyebrows and eyes that "burn more powerfully than fire, such that they could even look through the sun" (5931–5); he sports a snakeskin hat and clothes made from fish leather; his lower, equine half is absolutely black; and he can climb through mountains and valleys like a wild goat. On the other hand, Chiron has the appearance of a handsome old man, with exceptionally long grey hair and beard, but the rosy-cheeked complexion of a courtly gentleman; he has great understanding and can distinguish well between right and wrong; and he has thoroughly mastered such courtly accomplishments as playing the rote and harp, singing and playing chess. As such many noble and powerful kings trust him, and send their sons to his cave to be trained by him in "fighting and other skills". It becomes clear later on also that, at least in the case of Achilles, Chiron is a highly effective tutor: he forms the young man's character "as a seal forms wax when pressed into it" (6386–91), and as such Achilles does indeed turn out to be an excellent knight, made in his master's image.

In the two other thirteenth-century German adaptations of events surrounding the Trojan War it is, interestingly, the courtly, indeed artistic accomplishments of Chiron which are stressed. In Herbort von Fritzlar's *Liet von Troje* (c. 1200) Chiron is described as organizing a festivity for Peleus and the Nereid Thetis (17846–83) which is distinguished not just by outstanding catering, but by the excellent Muses, singers and poets whom he has invited – and who were appreciated much more, as Herbort rather predictably informs us, than comparably skilled artists are today. In the *Göttweiger Trojanerkrieg* (c. 1280), by contrast, it is Chiron the proficient harpist who comes to the fore, lightening by his accomplished playing the moods both of courtiers and of the Greek army encamped outside Troy (16669–78, 16832, 16847–9; Kern/Ebenbauer 171).

To the modern reader there is something almost Kafkaesque about the image of a wildly clad, horse-hooved centaur sitting at a chessboard or fingering a harp with virtuosic aplomb; and for certain one of the reasons for Konrad von Würzburg's lengthy treatment of Chiron is a delight in entertainingly imaginative, indeed eccentric storytelling. Konrad was nothing if not aware of the social and intellectual trends of his time, however; and there is no doubt that he is seeking also to propose Chiron as some kind of example for his audience to discuss and indeed to follow. The image of knighthood projected by the "good" centaur is, as Andreas Kraß says (96), fundamentally ambiguous – inevitably so, indeed, given the complex duality of Chiron's nature. Hence Chiron teaches Achilles, and as such perhaps also Konrad's readers, to be both bold, uncompromising fighters and polished, sensitive courtiers. Similarly he embodies certain chivalric characteristics (great power, high speed, the ability to negotiate difficult terrain, a viscerally animalistic yet controlled ferocity) which normally necessitate a horse and rider working together; yet at the same time he has qualities and abilities which normally exclude the equine half of a chivalric partnership (skill with weapons, tactical awareness, moral judgement, and a range of quintessentially aristocratic accomplishments). He reminds us, then, both of the importance of the horse's role in the chivalric scheme of things and of its essential subservience to the knight.

One final point needs to be made concerning both Konrad's *Trojanischer Krieg* and medieval constructions of chivalric identity more generally. Ulrich Barton (2009) has valuably observed that – unlike his Latin source – Konrad does not present Chiron's education of Achilles, for all its virtues, as sufficient in and of itself. Rather, before he can achieve any level of maturity, Achilles also needs a training in love – or, to be more precise, he needs the kind of direct experience of the joys and pain of love that so masculine, almost Spartan, a courtly tutor as Chiron cannot provide. For all their differences, this of course holds true also for all the other knights whose development and identities we have discussed. Their relationships with male horses, centaurs, lions or goats are essential to them in many ways; but to be truly "noble hearts", and truly integrated versions of themselves, Iwein needs his Laudine, Wigamur his Dulciflur, Gauriel his fairy queen, Willehalm his Gyburc, Bevis his Josian, and – last but not least – Achilles his Deidamia. Animals had many uses in helping knights to find and express who they were; but they could not do everything.

CHAPTER 5

Violence, Affection and Thirteenth-Century Animals

Abstract This chapter thematizes a tension common to both thirteenth and twenty-first centuries, namely that between violence and affection towards animals. Initially Francis of Assisi is presented as a (counter-intuitive) paradigm of this. Then other relevant examples are explored, including the barbaric treatment of cranes in falconry, two very different literary presentations of hunts, the bizarre use of judicial proceedings against animals, and the keeping of lapdogs by noble ladies. The analysis shows that violence and affection can be closely linked, and that both often reflect a quintessentially thirteenth-century instinct to challenge conventional understandings of the human-animal divide. The chapter ends with a postscript interpreting the replacement of parchment by paper in book production as spelling the end of a particular form of violence towards animals.

Keywords Hunting • Parchment • Pet-keeping • St Francis • Violence

I

Human beings' behaviour towards animals is, and probably always has been, characterized by a number of tensions and contradictions. Prominent among these is our tendency to treat our non-human fellow creatures with a strange mixture of affection and violence. This book, for example, has

© The Author(s) 2020
N. Harris, *The Thirteenth-Century Animal Turn*,
https://doi.org/10.1007/978-3-030-50661-2_5

been written by – and might well now be being read by – someone who is both a devoted pet owner and an inveterate meat eater. No doubt this is one of the commonest sets of double standards known to humankind: as Erica Fudge pithily puts it (10), "we rarely make the connection between the cat we live with and the cow we eat".

In the light of our own illogic it should come as no surprise to find that thirteenth-century people's thinking about the natural world was also often characterized by inconsistency and contradiction. A case in point here is Francis of Assisi (1181/2–1228). In popular understanding, of course, St Francis is regarded as one of the animal kingdom's best ever friends; and many scholars have shared this perspective. David Salter (25) reminds us, for example, of the eloquent and influential words of Lynn White, Jr., to the effect that Francis "tried to substitute the idea of the equality of all creatures, including man, for the idea of man's limitless rule of creation", and hence should be declared "a patron saint of ecologists" (1207). A similar thought seems to have inspired Pope Francis, upon his election, to take the name of "the man who loves and protects creation" (quoted in the *Catholic Telegraph*, 18th March 2013).

The fact is, though, that Francis's reputation as a benevolent revolutionary with boundless affection for the natural world is not really borne out by the evidence – especially if one concentrates, as for example Roger D. Sorrell has done, on early sources such as the *Vita Prima* and *Vita Secunda* of Thomas of Celano, rather than on later and more hagiographical writings such as the so-called *Little Flowers of St Francis.*

On the one hand, Francis's often compassionate, affectionate way with animals was undeniably endearing. When, for example, he encounters a cricket, coaxes it on to his finger and spends a delighted hour listening to it chirp, one can hardly avoid being drawn to him and wanting to emulate his childlike simplicity. Similarly, when he finds a hare caught in a trap, strokes it tenderly and then releases it (Thompson 70), one can begin to understand why he has become a powerful icon for many modern animal-lovers. Moreover such classic Franciscan motifs as his inviting baby robins to become friars or joining birds in song as they artlessly praise God (Sorrell 49) may be decidedly eccentric, but can only warm the heart of anyone who relishes the company of animals and thinks of them as equals.

Quite apart, however, from the fact that many of these patterns of pro-animal behaviour are not new, but rather reflect a Christian ascetic tradition established as long ago as the time of the Desert Fathers, there is another side to the story. Francis's view of nature often comes across as more in line with high-medieval hierarchical thinking than one might at first suppose. As Peter Dinzelbacher has pointed out (2000, 286), his response to animals often reflects less an interest in them in their own right than a desire to focus on what they might teach humans about God. Hence he picks up worms rather than treading on them because he is reminded of the expression associated with Christ in the Messianic Psalm 22, "I am a worm and no man"; and he loves lambs not because they are lovable in themselves, but because they are symbols of Christ. Moreover, if such perspectives are reminiscent of those of, say, Thomas of Chobham rather than Aristotle, then so are those actions on Francis's part which reveal his uncritical belief in the supremacy of humans over animals – in itself an attitude redolent, as Karl Steel (2011) has astutely shown, of both intellectual and structural violence.

One of the best examples of the somewhat inconsistent complexity of Francis's view of the natural world is the story of his taming of the wolf of Gubbio. David Salter (27–30) offers an acute analysis of this episode as related in the *Little Flowers* (or, more precisely, its Latin source). Francis, the story goes, is disturbed whilst staying in Gubbio to note that its inhabitants are so petrified of a ferocious, homicidal wolf that they will not venture beyond the city gate. So he takes it upon himself both to tame the wolf and to broker peace between it and its human neighbours. The former task is easily accomplished, by dint of Francis making the sign of the cross and ordering the wolf to change its ways, whereupon the latter bows its head and lies down at the great man's feet. Francis then addresses "Brother Wolf", roundly condemning it for its violent ways and ordering it to desist from them. The wolf makes various signs of accepting the saint's words, first nodding its head, then bowing and finally placing its paw in Francis's hand. This is the cue for Francis to close the bargain, promising on behalf of the good people of Gubbio that they will feed it daily for the rest of its life – as long, of course, as the wolf does not reoffend. So it all ends happily, with the wolf accompanying Francis back into Gubbio after the manner of a quiescent lamb.

One does not have to look very far beyond the warm glow that such a story gives, however, to realize that it has dimensions beyond conveying a sympathetic, benevolent, conciliatory side to Francis's character. He reasserts in no uncertain terms humankind's Edenic authority over animals; he accuses the wolf of depraved criminality for simply following its natural rapacious instincts; and he forces it to live both under the laws and within the physical environment of human civilization. Overall, his behaviour is a

far cry from that of a "patron saint of ecology" keen to assert the fundamental equality of all creatures; and, with regard specifically to the wolf, it
demonstrates violence every bit as much as it does affection.

Moreover Francis was well capable of encouraging or himself commit-
ting acts of unequivocal physical violence towards animals. Influenced perhaps by such New Testament stories as that of the Gadarene swine, he
plainly felt considerable animosity towards pigs. Whilst at the monastery
of San Verecundo, a sow attacked and killed a newborn lamb – a favoured
creature for Francis, as we have seen, in the light of its Christological
meaning. His reaction to the incident, however, was irrationally extreme,
cursing the sow and uttering an imprecation to the effect that no animal
should eat its flesh (Salter 45). Immediately thereafter, the sow died and
was thrown into the monastery moat – where it remained forever uneaten.
Nor was this an entirely isolated incident. Francis is also recorded as sup-
porting the actions of his associate Brother Juniper, who, having asked a
certain sick friar if he needed anything, was told that he wanted to eat a
pig's trotter. Hence "Juniper went to a wood where he knew that some
pigs would be feeding, selected an animal, and chopped off one of its feet
with a knife. Juniper then returned to the church of St Mary, where he
prepared the food, and gave it to his companion" (Salter 50). Self-evidently
the first of these stories can be read as a demonstration of God's justice
and power, and the second as a commendation of brotherly love. Seen
from the perspective of the pigs involved, however (and indeed from that
of their owners) they bespeak gratuitous and callous savagery, rather than
any qualities that might readily be associated with saintliness.

Francis was by no means not the only thirteenth-century figure who
can legitimately be accused of practising double standards in his dealings
with animals. Moreover there is evidence to suggest that the Animal Turn
we have been discussing exacerbated and sharpened the tension between
violent and affectionate urges, rather than alleviating them. On the one
hand, violence towards animals remained widespread, and is likely to have
increased given that the dividing line between animals and humans could
often seem disturbingly thin and porous: some people who felt challenged
or threatened by a heightened sense of animals' power and sophistication
no doubt felt moved to lash out at them in compensation and frustration.
On the other hand, that very awareness of the fragility and vulnerability of
the anthropological difference seems to have led people elsewhere –
including, perhaps surprisingly, in some hunting contexts – to draw closer
parallels than hitherto between the violence suffered by humans and that

visited upon animals; and at the same time affectionate closeness between humans and animals was also on the rise, as is exemplified particularly by a marked growth in pet keeping amongst aristocratic ladies. This in turn led some literary authors to employ pets as markers of female identity, and indeed sexuality. Even in contexts such as these, however, violence was seldom far away, be it committed against women, animals, or both; and overall, whilst the balance between violence towards animals and affection for them was slowly changing, there can be no doubt that, throughout the thirteenth century, it remained firmly weighted in favour of the former.

II

Our survey of these trends must begin by considering a notable recent treatment of medieval violence towards animals, Karl Steel's *How to Make a Human*. This study is of particular relevance to our current discussion because of Steel's claim that human violence towards animals can often be compensatory in nature, a kind of reflex action that affords temporary release from feelings of insecurity, challenge or threat – such as that frequently made in the thirteenth century to time-honoured conceptions of the anthropological difference. Hence violence towards animals can appeal to people as a means of seeking to assert their "prejudice of 'natural' human ascendency over other animals" (3) and of attempting to claim "a unique, oppositional identity for themselves" (15). On this basis Steel discusses a wide range of relevant cultural manifestations from the medieval period: theological discourses that posit an intrinsic link between humans' rationality and their perceived calling to subjugate animals; the denial to animals of any possibility of eternal salvation; notions of the body that assert a fundamental prejudice in favour of bipedality and the possession of hands; attempts to legitimize the use of force against animals whilst at the same time illegitimizing any self-defence or self-determination on their part; questionable processes of domestication and ostracization; and, not least, various forms of butchery directed at pigs and boars.

   Steel's emphasis is often on what one might call the indirect violence implicit in certain thought patterns; but the consequences of these for direct physical aggression are plain. Medieval cats, for example, were often singled out for acts of casual brutality. They were regularly skinned for their fur, a practice to which even Bartholomaeus Anglicus refers in his encyclopaedia. Moreover Jacques de Vitry has a story (Crane no. 8) describing a dice game at the University of Paris in which a cat is forced to "throw" a die; if it throws a higher number than the one a student has just thrown, it is fed and allowed to survive into the next round; but if it throws a lower number, it is skinned and its skin sold – "no doubt to provide the students with funds for other varieties of dissolution" (Jones 108).

Grievous acts of violence towards animals were indeed seemingly so pervasive as to be routinely sanctioned even by those thirteenth-century figures whose attitudes on the matter might be expected to be more humane and forward-looking. A particularly good example is again Frederick II. However notable an ornithologist, menagerie-keeper and indeed animal-lover, Frederick was plainly routinely guilty of barbaric violence towards birds in his care, as well as of some quite startling double standards. In his *De arte venandi cum avibus* he is entirely open about such things. On the one hand, Frederick's readers are enjoined to take great care of their birds: "it is imperative that he [the falconer] should also be governed in his relations with his birds by the state of their health" (157); falcons should be given regular baths, "not only during the period of taming our falcons but as long as we own them" (191); and immense care must be taken when capturing wild falcons to ensure they are not injured in the process (144). The reasons Frederick proceeds to give as to why this is important are, however, only too revealing:

> This consideration holds out the only hope for good results, because those birds that are injured when trapped, either by their own violent efforts to escape or when they are roughly released from a net, rarely if ever recover from such ill-usage; either they die from their injuries or, for a long time, if not always, remain perfectly useless (trans. Wood/Fyfe 144).

We are in the presence, then, not so much of any charitable or sentimental attachment to the birds involved, but of ruthlessly hard-headed business sense. This point is reinforced still more strongly when Frederick speaks of cranes, and the roles they can play in the training of falcons. Cranes can be used as live lures; and, if they are, the routine indignities and cruelties to which they are subjected betoken a frighteningly pitiless and cold-hearted approach that really has to be read to be believed:

> Procure a live crane; and, since that species defends itself with its beak and feet,… its claws must first be coped and its beak bound to prevent it from injuring the gerfalcon. In coping, the crane's claws are charred by means of a lighted wooden splinter until their sharp (needle-like) points are blunted

and thickened... The beak is rendered harmless by passing a small cord through the opening in the two nostrils and binding the lower mandible to the upper, so that the crane cannot bite. Then the crane must be seeled so as to render her quite blind and unable to see when and where to strike the falcon (257).

If the crane is seen as potentially too strong for this task, however, it is then to be subjected to yet more violence:

Two sticks are sewn one into each end of a cloth, and this fabric is used to enfold the crane, whose feet and legs are bent beneath its body in the sling. The sticks are brought together over the back of the crane from the tail to the shoulders, near the neck. The two upper extremities of the poles are then bound together by one end of a cord whose opposite end unites their lower extremities near the crane's tail. In this fashion the bird may be slung by the cord placed for that purpose around the carrier's neck... The crane's efforts to escape, the carrying about, and the bending of its legs in the sling, which causes pain in the knees, all contribute to stupefy and weaken the bird to a point where it can be utilized in the train (257f.).

Finally it is clear that such viciousness was at times visited on cranes even after their death. A skilled falconer can, it seems, convincingly reproduce the sound of a live crane's voice by working on the larynx of a dead one:

As it is an advantage for the falcon to recognize the cries of the crane... the falconer should lay bare a crane's larynx as close to the mouth as possible, and remove its heart. No other incision should be made in any part of the body, except that as the larynx is to be drawn out after the first cut is made, the skin of the crane should be slit a little downward from the opening (at the throat). The falconer should then grasp the end of the pulmonary tube and blow into it, inflating the lungs and trachea with air. Taking care not to allow the air to escape between his fingers, let him pinch the end of the larynx, near his mouth, and remove it from his lips. When he wishes to imitate the voice of the live crane, he has only to compress the sides of the bird and release the end of the tube held in his fingers. The crane will then emit the same call as a live one (261).

III

It is no doubt significant that Frederick's highly questionable methods of dealing with cranes should have been practised in a venatorial context. Certainly hunting provided, throughout the Middle Ages but especially during its courtly centuries, a centrally important realm in which aristocrats could demonstrate, often violently, their superiority not only over the animal world but also over "lesser" human beings. The sometimes forcible exclusion of peasants and others from protected aristocratic hunting grounds – itself a clear case of structural violence – is not really thematized in courtly literature; but the dominance of the superior huntsman over both his animal quarry and over his less courtly colleagues certainly is.

A good example of this is the extensive hunting scene in Gottfried von Strassburg's *Tristan* (c. 1210). The narrative in question begins at a late stage of the process: a stag has been cornered and is killed by hunters belonging to King Mark of Cornwall – a scene which happens to be observed by the young Tristan, who has not long escaped from his abduction by Norwegian pirates and is passing himself off as either a lost local boy or the son of a noble merchant. It is not long, however, before his innate regal courtliness shows itself: he is, to say the least, unimpressed by the chief huntsman's attempts to lay the dead stag out on the grass "on all fours like a boar" (2791; trans. Hatto 78), and offers to demonstrate a superior way of skinning, excoriating and eviscerating the stag, and then presenting it – antlers to the fore – to the members of court. He proceeds to do all this in hugely impressive style. Both Tristan's knowledge and his sophisticated techniques are seen by those present as combining with all the other aspects of his persona to create an impression of the most refined courtliness: "They inwardly considered his bearing and behaviour, and it pleased them so much that they delighted to watch it. They confessed to themselves that everything about him was noble, his clothes rare and magnificent and his figure of perfect build" (2852–9; Hatto 79). For the moment at least, Tristan's evident difference from, and superiority over both the stag and the Cornish huntsmen is both effortless and unproblematic; but it is worth recalling that they are founded in the first instance on what amounts to a set of advanced butchery skills, used on an animal who has just been killed, and will before long be eaten, by his new associates at the Cornish court.

In this scene, as in most other descriptions of hunts in thirteenth-century literature, there is no hint of the violence meted out to the hunted animal being reflected in, or equated with, violence perpetrated on human beings. Other contemporary contexts, however, contain signs that people were beginning to draw more parallels, and becoming more aware of a continuum, between these two superficially unrelated phenomena. In a world preoccupied by an apparently narrowing gap between the human and the animal, this no doubt represented one side of a coin whose other, uglier face depicted some of the forms of insecure, compensatory violence alluded to above.

A particularly fascinating example of brutal violence being directed at man and beast alike occurs in a pivotal episode of the anonymous *Nibelungenlied* – an epic poem whose ethos is very different from that of Gottfried's *Tristan,* but which was composed (at least in the form we now know it) at around the same time. Here the heroic Siegfried is very much hunter and hunted, killer and killed. In a scene beloved of some of the most notorious figures in twentieth-century German public life, Siegfried is "stabbed in the back" by the villainous Hagen whilst drinking at a forest spring. More precisely, Hagen spears him between his shoulder blades at the sole spot left vulnerable after Siegfried's bath, years earlier, in the blood of a dragon he has slain. As such, the act arguably involves an element of posthumous revenge on the part of the dead creature.

The extended episode describing the hunt occupies almost all of the so-called sixteenth adventure of the *Nibelungenlied,* and features a varied and substantial cast of animals. The first of these are the two boars of which Siegfried's wife Kriemhild dreams on the night before his murder: "I dreamt last night, to my grief, how two wild boars chased you over the heath, and flowers grew red there... I am sorely afraid of some conspiracy or other" (921,2f., 922,1; trans. Edwards 87). No connection is drawn by the narrator between these boars and any specific characters; but by this stage of the story any mildly attentive hearer or reader will have no difficulty in associating them with King Gunther and his powerful kinsman Hagen – whom we know to be engaged in just such a design on Siegfried's life. For his part the latter, as one would expect from the protagonist of a heroic epic, pays no heed to his wife's warnings; but her fears are subsequently proved only too justified.

The events of the hunting expedition in which Siegfried and the Burgundian court participate unfold in three stages. First we see Siegfried as champion hunter. Left alone save for a single, if highly effective tracking-dog, he makes short work of a lion, a bison, an elk, assorted harts and hinds, "four mighty auerochses and a fierce buck" (937,2; Edwards 88).

Moreover his prey also includes two fine boars, a motif which surely antici-
pates the eventual deaths of Gunther and Hagen, appropriately enough at
Kriemhild's hands or behest, at the end of the epic. There is a sense indeed
in which the prodigal loss of animal lives occasioned by Siegfried is presented as the result of violent murder, as much as of courtly hunting tac-
tics. He "slays" the creatures, much as he would human opponents, with
his bare hands, bow and arrow or sword, such that "little could evade
him" and "many of the beasts had to forfeit their lives there and then"
(937,4, 942,1; Edwards 88f.). All this seems some way removed from the
painstaking mannerliness and method of Tristan's venatorial world, and
rather to blur the lines between these animal victims and potential
human ones.

If anything still more unusual is the second section focusing on
Siegfried's hunting exploits, namely his dealings with a mighty wild bear.
Even after having decimated much of the forest's livestock, Siegfried
retains sufficient energy to ride down this bear, and then pursue it on foot,
capture it, and tie it to his horse's saddle. In this way he brings it back to
his fellow hunters' camp. Once there, Siegfried resolves to have his fun
with it, causing it to rampage destructively through the camp kitchen and
generally to terrify both hounds and hunters. When the bear eventually
flees, we are told that, of course, "none could keep up with it except
Kriemhild's husband. He caught up with it and then slew it with his sword.
After that they carried the bear back to the fireside" (962,2–4; Edwards 90).

This episode with the bear is far from easy to interpret. D. G. Mowatt
and Hugh Sacker, for example, do so imaginatively if somewhat
bet-hedgingly:

> In one sense the bear is Brünnhilde, caught by Sifrid, set loose in Burgundian
> society, captured a second time by Sifrid, and finally rendered harmless... In
> another sense, the bear is Sifrid himself, an uncomfortable guest in
> Burgundian society, a well-meaning disaster (the bear is only trying to run
> away), and finally a ritual murder victim. And in so far as Sifrid and
> Brünnhilde are one composite symbol, the bear is the spark in their relation-
> ship that Sifrid stamped out (89).

This interpretation certainly takes appropriate account of key aspects of
the epic (such as the implication that Siegfried and the initially Amazonian
Brünnhilde are in some symbiotic way destined for each other); and it
points the reader to several possible links between human and animal

victims of violence. In the end, though, it is surely both far-fetched and, in that it seeks to associate the bear on a one-to-one basis with individual characters, unduly reductive. In general terms one suspects that the sequence of events involving the bear appealed to the *Nibelungenlied* poet as an entertaining delaying device that might both contrast with and build up suspense before the climactic murder of Siegfried. After all, it contains elements of both bear-baiting and bear-dancing, demeaning pursuits which will nevertheless have been known to an early thirteenth-century audience and have been regarded by them as legitimate, diverting forms of sport.

An interpretation of the bear which arguably does justice both to the poem's narrative structure and to its complex amalgam of elements drawn from both courtly and pre-courtly cultures is, however, to see it as representing the pinnacle of the animal kingdom, a powerful enemy which only the finest hunter/knight, Siegfried, can take on and defeat. In narrative terms the bear is presented as Siegfried's ultimate and most difficult animal opponent; and in the audience's eyes this will no doubt have seemed appropriate, given the bear's historic role as king of the beasts. According to Michel Pastoureau (2007), it was only between around 1000 and 1200 that, in the light of a concerted, long-term campaign on the part of the Church, the bear had finally to cede its pre-eminent position to the lion; but traces of earlier thinking on the matter remained; and these will for certain have been familiar to readers/hearers of a work such as the *Nibelungenlied*, which predicates some knowledge on its audience's part of older Germanic traditions. Moreover such an audience will for certain have registered a marked contrast between the ease and speed with which, early in the piece, Siegfried has dispatched the lion, and the considerable trouble he has with the bear. Seen in this light, Siegfried's conquest of the bear constitutes the height of his career vis-à-vis the animal kingdom: he emerges as both fundamentally different from and effortlessly superior to even its most illustrious and formidable representative. All the more humbling and tragic, then, does the following scene appear – in which Siegfried is first chased and cornered (however unknowingly) by Gunther and Hagen, and then attacked from behind by the latter, as if he were an animal refreshing itself at a watering hole. The pre-eminent scourge of the animal kingdom himself becomes, in death, the victim of the kind of violence standardly meted out – by the likes of him – to members of that kingdom.

In at least one other thirteenth-century context, albeit a considerably later and fundamentally different one, there were signs of animals being treated essentially as human beings when it came to matters of violence and punishment. This can be seen in the use of judicial means to chastise individual animals or groups of animals held to be guilty of crimes against people or their property – never, it must be stressed, against other animals. Courts to try animals were convened on an at least occasional basis from the thirteenth century onwards, especially in francophone areas and in the West of the Holy Roman Empire. There is no evidence to suggest that they ever occurred with anything like the frequency of witch trials, but it is interesting and no doubt revealing that their temporal and geographical concentration is not dissimilar to these.

The first recorded instance of an animal court took place at Fontenay-aux-Roses near Paris, in either 1266 or 1268. A pig was accused of eating a child, found guilty and sentenced to death by burning. In the decades and centuries that followed domestic animals like pigs, horses and oxen tended to be the main objects of such procedures in criminal courts; meanwhile ecclesiastical courts dealt with damage done to life and limb by wild animals, often for example swarms of insects who destroyed large quantities of crops. In both contexts the law's full panoply was brought to bear: judges, lawyers, bailiffs and hangmen were all involved, and all the usual expense incurred. Moreover the animal miscreants themselves were often treated remarkably like people. This seems to have applied particularly to pigs: in 1386 a sow was dressed in man's clothes in preparation for her public hanging; in 1394 a pig is recorded as having been kept in a human jail and served standard human rations; and in 1457 six piglets were pardoned on grounds that we would now describe as mitigating circumstances. They and their mother had been jointly accused of killing a five-year-old boy, but the sow was hanged and her young pardoned, "both for their youth and for the bad example set by their mother" (Salisbury 112).

The reasons for such behaviour on the part of highly trained and responsible people are hard to fathom. Much of it makes no sense at all. Certainly no serious theologian would ever have argued that animals such as pigs and oxen were possessed of sufficient reason or moral conscience to be legitimately held responsible for their own destructive actions. Similarly, the relevant legal processes were all predicated on the plainly absurd assumption that animals were able to understand and respond to human speech; and the church courts' sanction of excommunication was not infrequently imposed on creatures who by definition could never have been in communion with the Church in the first place.

On the surface, then, those who organized and prosecuted such trials were operating on a basis that palpably subverted their capacity for common sense reasoning. Perhaps they were acting in a kind of "crisis mode", deeply disturbed by a real and present danger against which they felt powerless, and in consequence feeling the need to reassert their dominance over animals by lashing out at them with a darkly visceral destructiveness. Alternatively, one might argue with Peter Dinzelbacher (2002, 406), that prosecutions of animals were by way of "show trials" consciously designed to assuage the fears of an uneducated public:

> Animal trials took place only under extremely unusual circumstances in order to help the local community cope with an otherwise recalcitrant threat – not because they were proven to work but because they created the impression that the authorities were assiduously maintaining law and order in a cooperative and decided manner, even if the delinquents were not human beings.

Whichever of these scenarios is closer to the mark, there can be little doubt that animal trials resulted from fear, whether of animals themselves or of what they were seen to represent; but they could only have taken place at all in a context where the whole nature of animals and their relationships to humans were already being rethought and renegotiated. They undoubtedly betray a "tendency to reduce the ontological distance between man and beast" (Dinzelbacher 2002, 420); and to that extent, in spite of their disturbing strangeness, they are in line with many of the other thirteenth-century developments we have been discussing.

IV

In a chapter which sets out to discuss evidence of both violence and affection as they are expressed towards thirteenth-century animals, we have hitherto focused largely on the former. The balance can now be redressed to some degree by introducing certain sources which document the existence of a burgeoning counter-tendency for people to draw close to animals, to treat them in a privileged way and to show them, in some cases, some quite demonstrative levels of affection.

This rise in affection for animals is clearly reflected in the close literary relationships between knights, horses and other companion animals we discussed in the last chapter; and it can also be seen in what seems to have been quite a remarkable growth in pet keeping from the late twelfth century onwards. Not that there was much in the way of ecclesiastical sanction for such an activity: Thomas Aquinas denied the very possibility that a human being could love an animal other than in a metaphorical sense (Dinzelbacher 2000, 288); and many other critics, particularly clerical ones, "objected to ostentatious pet keeping, viewing it both as an extravagance and a distraction from one's duties and responsibilities, in particular charity to the poor" (Walker-Meikle 18). Even Franciscans were banned from keeping pets by their Rule of 1221. Nevertheless many people did keep pets – primarily, it seems, clerics and noble ladies. This is doubtless a reflection of the essentially indoor lifestyle of these two groups, which necessitated a very different kind of companion animal from the horses, hunting dogs and falcons favoured by secular lords.

By definition it is very difficult to reconstruct the nature of medieval people's affection towards their pets, or the ways in which they expressed it – though one way which is often criticized in literary and other contexts is chronic overfeeding. Stephen of Bourbon, for instance (I, 191), has an exemplum about the damage an over-indulgent owner has done to the health of a massively overweight dog; and, rather later, one of the various moral weaknesses attributed to Prioress Eglantyne in the *Canterbury Tales* is that she feeds her little dogs inappropriately on "roasted flesh, or milk and fine white bread" (Prologue, 146–50). Other potential indicators of affection are the proliferation of small dogs depicted on ladies' seals and funeral effigies, the survival of numerous elaborate cushions intended for pet use, and – especially towards the end of the Middle Ages – poetic outpourings of grief over the death of pets (Walker-Meikle 31–8, 50f., 75–81). How far manifestations such as these betoken genuine intimacy or heartfelt emotion, and how far they reflect pets' role as identity markers analogous to, say, the heraldic emblems of battles and tournaments is, of course, ultimately impossible to say.

Whatever the precise contours of the social background, however, high-medieval literature features pets of various kinds. Most of these, inevitably, are in the possession of religious males or, especially, courtly females. For their part, literary reports of companion animals kept by priests, monks or indeed saints go back a long way – one thinks of St Jerome and his lion, or the delightful eighth- or ninth-century Irish poem about the relationship between an Irish monk and his cat Pangur Bán, somehow together and somehow apart as they perform their separate but compatible daily tasks. There is not much evidence to suggest that the thirteenth century developed this tradition with any consistent assiduity, but there are certainly some works from the period which contain animals kept by the likes of saints and patriarchs; and these tend to be treated in a basically comic way that suggests a desire to play with, or even parody what one might call the Pangur Bán tradition. In the *Weltchronik* (world chronicle) of the Viennese patrician Jansen Enikel (probably 1270s), for example, the patriarch Noah seems to have a pet goat. Certainly such an animal is described as running one day alongside Noah when the two of them find some vines in a forest. The goat proceeds to eat its fill of the grapes on offer, such that it becomes drunk and wine runs down its face over its beard (2815–22). This "clever" (2807) if inebriated goat is briefly celebrated as the discoverer of wine and its many benefits, but thereafter – we assume – returns to anonymous domesticity somewhere in Noah's household.

A much greater role is meanwhile assigned to a raven which has been kept for twelve years (361f.) by the saint-king Oswald, as recounted in the curious – and ultimately undatable – cross between saint's life and bridal-quest epic that is the so-called *Münchner Oswald*. This raven, whom God endows with speech to enable him to fulfil an important role as Oswald's messenger to his would-be bride, is in many respects a comic character. In the various expansions and digressions about the raven in which the narrator indulges, the bird's marked verbosity seems to express an entertaining keenness to use to the full the gift of speech he has just been given; and he is also repeatedly and divertingly guilty of two other very human weaknesses, gluttony and vanity. At the same time, however, Oswald's pet raven invites serious reflection on both the openness and the restrictiveness of the human and animal kingdoms. He is palpably a member of the latter; but he is a bird who both speaks and thinks like a human; who considers himself a cut above other non-human creatures; and who manifests no real desire or ability to communicate with them. Amongst humans meanwhile the raven enjoys a certain privileged freedom precisely because he himself is not human; but this freedom and the privileges associated with it are conferred by humans (and indeed by God) only for particular purposes and at particular times. In other words, whilst actively performing the – normally unequivocally human – role of Oswald's messenger, the raven consorts with the king and his princely equals, and is allowed to converse with them with a familiarity and outspokenness that would be denied to the vast majority of their subjects; but as soon as his job is done, he is quite literally forgotten about. As such he reminds us that it is possible, not least for the privileged domesticated animal, to inhabit both the human and the animal worlds without being securely part of either.

But what of animals whom it might be easier for us to identify as pets in the modern sense? On the whole medieval cats tended to be viewed simply as working mousers and occasional sources of fur. Lapdogs, however, became especially popular with ladies in late twelfth- and early thirteenth-century Italy, and plainly retained this role and status for many centuries thereafter. Already in the thirteenth century, indeed, lapdogs became so widespread, and so closely identified with their female owners, that the possession of one was frequently seen as an identity marker especially for noblewomen – so much so that, "by being part of its owner's everyday life, sharing in many activities, the pet became part of the owner's persona" (Walker-Meikle 3).

The most famous thirteenth-century literary lapdog is Petitcrieu in Gottfried von Strassburg's *Tristan* (c. 1210). Petitcrieu has been given to Duke Gilan of Swales "by a goddess from the fairy land of Avalon as a token of love and affection" (15810–14; Hatto 249), and is now the Duke's "heart's delight and balm to his eyes" (15802f.; Hatto 249). Gilan notes that the dog also has the capacity to cheer the melancholy Tristan, by dint of its two magical features: it is astonishingly multicoloured, such that no one can tell with any certainty what colour it actually is (15822–5); and, on a chain of gold around its neck, hangs "a bell so sweet and clear that, as soon as it began to tremble, melancholy Tristan sat there rid of the sorrows of his attachment and unmindful of his suffering for Isolde. The tinkling of the bell was so sweet that none could hear it without its banishing his cares and putting an end to his pain" (15851–63; Hatto 250).

The upshot of this is that, when Tristan has helped Gilan by seeing off the threat of the giant Urgan, he asks to be given Petitcrieu as a reward. Upon delighted receipt of the dog, Tristan immediately arranges for him to be given to Isolde, in order to restore happiness to her (16280–2) – albeit, of course, at the cost of his own. The plan does not really work, however: even though Isolde proves to be the perfect early thirteenth-century pet owner – giving Petitcrieu "a delightful little kennel of gold and precious things" and a rich brocade to lie on, and never letting him out of her sight (16341–55; Hatto 256) – she can take no comfort from him. Before long she reaches crisis point:

As soon as the faithful Queen had received the dog and heard the bell which made her forget her sorrow, she had reflected that her friend Tristan bore a load of troubles for her sake, and she immediately thought to herself: "O faithless woman, how can I be glad? Why am I happy for any time at all while Tristan, who has surrendered his life and joy to sorrow for my sake, is sad because of me? ... He has no life but me. Should I now be living without him, happily and pleasantly, while he is pining? May the good God forbid that I should ever rejoice away from him!" So saying, she broke off the bell, leaving the chain round the little dog's neck. From this the bell lost its whole virtue... But this meant nothing to Isolde; she did not wish to be happy. This constant, faithful lover had surrendered her life and joy to the sadness of love and to Tristan (16362–406; Hatto 256).

With his trademark exquisite craftmanship, Gottfried has seen to it that the Petitcrieu episode as a whole, and the above passage in particular, culls the very essence of Tristan-love. This brings both incomparable joy and incomparable sorrow; it is selfless and self-sacrificing; it is absolute and all-consuming; it has indistinct but ever-present overtones both of Christianity and of magic; and, last but not least, it is utterly doomed. Isolde does not wish to be happy because she knows she never *can* be happy. Her fate is love, Tristan, and nothing else. And this heightened, implicitly supernatural but ultimately destructive quality of Tristan-love means that Petitcrieu, for his part, cannot be "just" a normal dog. If he is effectively to construct the identity of Tristan and Isolde as lovers, he too must possess powers and embody possibilities that transcend standard-issue dog-hood; but these powers must also end badly. In other words, his bell must break. Tristan-love always leaves destruction in its wake.

V

In spite of his uniqueness, Petitcrieu is typical of thirteenth-century literary lapdogs in being closely associated both with his lady's love life and with an act of violence. Elsewhere, however, this violence tends to be considerably more destructive than simply wresting a magic bell from a dog's collar – and it can affect animals, human beings, or both.

The short romance *La Chastelaine de Vergi* (mid-thirteenth century), for example, has a baleful conclusion for the heroine and several other human figures – though, as far as we know, the *chastelaine's* pet dog survives. He is in essence a go-between: when summoned by the *chastelaine*, her secret lover is instructed to wait in an orchard, and only to venture forth from it into her chamber when the dog crosses the orchard, indicating that the coast is clear. The dog performs this function several times, but his role is not otherwise developed until shortly before the end. The *chastelaine's* rival in love the Duchess of Vergi (also her aunt by marriage) uses the dog as a euphemism for the *chastelaine's* illicit affair when cruelly remarking to the latter that she has been "a clever mistress to have learnt how to train the little dog" (90). Having learnt in this way that her lover has broken his promise to keep their liaison secret, the *chastelaine* dies in despair, the knight finds her body and kills himself, and the Duke finds both bodies before exacting a particular brutal form of revenge by killing his wife, the Duchess. As Walker-Meikle (93) has pointed out, however, in subsequent illustrations or depictions of the tale, "the dog is always a key component, witnessing the lovers' pledge, sitting with his mistress, being put out into the garden, present at the lovers' embrace and at their final discovery". More so than Petitcrieu, then, the *chastelaine's* pet is a consistent, indeed almost ubiquitous emblem of a doomed love.

In popular comic tales also the link between women, their animals and violence can be a disturbingly close one. Kathleen Walker-Meikle (60) refers us to a *fabliau* in a thirteenth-century manuscript from Dijon which has a decidedly tragicomic plot: a lady is sharing her bed with her pet dog, when her husband returns, perceives a shape in the bed next to his wife, assumes it is her lover and stabs her. Moreover violence in the form of what can be called at best semi-consensual sex occurs in several comic tales featuring pet animals and naive young women. A relatively mild example of this is *Der Sperber (The Sparrowhawk)*, a German tale datable to the first half of the thirteenth century. Here the innocent ingénue is a young nun. Walking in the vicinity of her cloister, she encounters a handsome knight with a pet sparrowhawk on his hand. She takes a fancy to the bird, says so, and the knight offers to sell it to her in exchange for her love ("minne", *Novellistik* 125). The nun, of course, has no idea what such a currency might consist of; but, in the seclusion of a nearby orchard, she very soon finds out – and indeed discovers that the experience "causes her the pleasantest of pains" (168). So pleasant are they indeed, that the girl swiftly and proactively offers to pay the knight two further instalments of *minne*, an offer he is only too willing to accept. For the modern reader there are of course decidedly disturbing overtones in all this, with regard both to the commodification of love and to the girl's inability fully to understand what she is doing; and this discomfort is intensified when we read of the sparrowhawk (a clear symbol of the girl's sexuality) being tied for the duration to the branch of a tree (163–5), and of the girl's Mother Superior subsequently pulling out her charge's hair and beating her to within an inch of her life (230–4). For the narrator, however, such violence is clearly just part and parcel of a comic tale, and he draws things together quickly enough to provide us with a happy ending: next day the nun again meets the knight and asks him to pay back the *minne* he has received from her, in order that she might regain something the Mother Superior has referred to as her *magetuom* ("virginity", 268); he, again, can hardly believe his luck, and gladly repays all three instalments to the full; the girl is happy that she has now got her virginity back; and even the Mother Superior concludes that she has no option but to shrug her shoulders and get on with life.

The only character whose fate is unknown is the sparrowhawk: it is clear that the nun has intended for the knight to have it back (263), but far from clear that he actually takes it, or indeed that the nun herself retains any interest in it. So the reader is left unsure of precisely how to interpret the bird: if it remained with her, then presumably we would see it as a symbol of her (actually) lost virginity; but if it returned to the knight, then it would arguably represent a sexuality that is first dormant, then briefly awakened and now dormant again. It may well be, of course, that the poet himself simply did not know what to make of the motif of a sparrowhawk, like all birds of prey generally associated with vigorous masculinity, becoming the pet of an innocent young woman; and certainly it is not surprising that, in a later, expanded German reworking of the story *(Das Häslein)* the woman is given a rabbit.

At least we can assume that, one way or another, the eponymous bird of *Der Sperber* survives the storytelling process; that, however, cannot be said of the three animals who feature in Sibote's *Frauenzucht*, a version of the "taming of the shrew" story believed to have originated at the court of Frederick II's son Manfred (c. 1232–66). A young knight is determined to marry a beautiful but ferociously disobedient young lady, and – between their betrothal and their marriage – resolves to "train" her (hence the story's title). For this purpose he uses three quintessentially aristocratic animals, his sparrowhawk, his dog and his horse, all of which accompany him and the lady on a journey down a (no doubt symbolically significant) narrow path (270). Towards all three animals the knight behaves with barbaric intolerance. When the hawk seeks to fly from his hand in pursuit of a passing crow, he wrings its neck like that of a chicken (273–87); when the dog pulls at its leash, the knight cuts it in two with his sword (297–301); and when the horse is reluctant to be spurred on, he decapitates it (308–17). The slaughter of his horse means, of course, that the knight no longer has a mount; and so, in a reversal of the Aristotle and Phyllis motif, the lady has to step into the breach. In spite of what has happened, she clearly retains spirit enough to ask him to ride her without a saddle (335–7); but when he refuses, and equips her with both saddle and bridle, she complies readily enough. After nearly a mile riding on her back, the knight relents, knowing that he has won; and she proceeds to make him "the best wife ever…, respecting his will at all times" (383f., 387).

This story too, then, has a happy ending of a sort – especially given that, shortly before its conclusion, the knight succeeds also in taming his mother-in-law, at whose hands – and fists – his father-in-law has suffered over some thirty years. Such marital harmony as accrues, however, is bought at a huge cost, both to the women and the animals involved. In stories like *Der Sperber, Frauenzucht* and many others the former are routinely objectified, intimidated, dehumanized and subjected to the combination of physical and structural violence endemic in so many patriarchal societies. And in this process animals are also often victims – used as means to an end, exploited and sometimes killed in ways that evince precious little concern for the value of their lives.

VI

All the more reason, then, to end this chapter with an essentially upbeat postscript. At the same time as pigs were being sentenced to death in France and Germany and animals of many kinds were being slaughtered throughout Europe, technological developments were underway in Italy that spelled the beginning of the end for one particularly acute form of cultural violence – namely the mass killing of animals to make parchment, and hence books.

The first paper mills in medieval Europe were opened around 1235 in Fabriano, in the province of Ancona. The ancient art of papermaking, which had made a long and slow progress from China to Italy via North Africa and Spain, was fruitfully combined there with new technologies of metal-processing and weaving (Müller 44). This meant that in Fabriano and, from around 1255, other centres such as Milan and Genoa (Hills 38), all the phases of papermaking were revised and modernized – from pulp-beating through the formation of leaves, to drying, smoothing and liming. These and other innovations (notably the increased use of water-power) made paper more competitive, with the result that by around the mid-fourteenth century, manuscripts made from it outnumbered those still being made using parchment.

Paper was not uniformly welcomed, not least due to early concerns about its quality and durability – Frederick II went so far as to prohibit its use for public documents (Hills 38). Whatever its pros and cons from a human point of view, however, for animals the increased use of paper was all gain. Given that the skin of one calf was standardly needed to make two sheets of parchment, one can assume that, say, the 30 parchment copies of Gutenberg's Bible (641 leaves) will have cost the lives of some 9630 animals used as writing surfaces. In respect of the 180 copies Gutenberg printed on paper, however, the death toll will have been limited only to the animals whose leather provided the covers and those whose skins and bones were used to make the necessary gelatine; and the latter normally came from animals already killed by tanners anyway (Hills 44).

From a human point of view, one can certainly argue that reading about animals in a paper manuscript is a less satisfying, or at least less complex experience than reading about them in a book constructed from their skins. Sarah Kay (2017) has observed this with specific regard to parchment bestiary manuscripts (above all illustrated ones). The act of reading these, she says, "can create a feedback loop between the page as an animal surface and the texts written on it" (4); after Slavoi Žižek, she calls this process a "suture", in which "the distinction of levels between content and medium on which reading normally relies is momentarily suspended, with uncanny effect" (5). Moreover, "given that the content aims to instruct humans while the medium resembles human skin, this suture potentially implicates the human reader too, whose bodily surface is likewise caught up in the collapse of distinctions between levels whose difference from one another is usually taken for granted" (5).

On the basis of these perceptions, Kay goes on to demonstrate various ways in which "the page as such intervenes" (20) in the reading process. Parchment is revealed as making possible various connections: between the "book" of nature, the book of scripture and the tunic skins donned by human beings since the Fall; between the holes and tears on the parchment's surface and the bodily orifices and wounds of both people and animals; or between the putative inner lives of both animals subjects and human readers. Self-evidently paper pages cannot achieve such things, or at best can do so only in an indirect and impoverished way. When paper books about animals replace parchment ones, there is no doubt the reading experience loses richness and suffers a qualitative loss.

On the other hand, once paper manuscripts became – by medieval standards – mass-produced items, this qualitative loss was offset by a quantitative gain. Books became much easier and cheaper to produce, and hence many more people were able to read about animals, and in many different places, than had been the case previously. One can see this to varying degrees in the manuscript traditions of the encyclopaedias by Bartholomaeus Anglicus and Thomas of Cantimpré, which have been assiduously studied by Heinz Meyer (2000) and Baudouin van den Abeele (2008) respectively. In Thomas's case more than half of the surviving 222 manuscripts are paper ones from the fourteenth or fifteenth centuries, and they facilitated a much travelled manuscript transmission which resulted in Thomas being read from Spain to Switzerland and from Ireland to Romania (van den Abeele 2008, 151). In the case of Bartholomaeus's *De proprietatibus rerum* paper was particularly important as the medium on which no fewer than 52 editions were printed between 1470 and 1607 (Meyer 240). That said, only around 40 of the 188 manuscripts Meyer describes were written on paper, the rest on parchment. This is a much smaller proportion than one would normally expect; and very unusually, there are even 13 parchment manuscripts from the fifteenth century. We shall never know why, but it is certainly tempting to wonder whether some late-medieval readers of nature encyclopaedias really did retain a sense that a book *about* animals was best written *on* animals – at however catastrophic a cost to the local veal population.

CHAPTER 6

Conclusion

Abstract This short conclusion summarizes the developments analysed in the book and assesses how far they remained influential beyond the thirteenth century. It argues that several of them retained considerable significance well into the early modern period, notably the authority and methodology of Aristotle, the sometimes tortured questioning of the animal-human divide, and many aspects of the use of animals for didactic purposes. By contrast, the medieval chivalric code had largely perished well before 1500 – though heraldic devices remain widespread in many contexts; and the tension between violence and affection towards animals is still very much with us today.

Keywords Aristotle • Covid-19 • Emblem-books • Modern heraldry • Montaigne

I

The foregoing chapters have argued that thirteenth-century Western Europe witnessed an Animal Turn – understood as a cultural change in which scholars become increasingly aware of and interested in animals, and develop new ways of looking at them and writing about them which have implications also for sectors of society situated outside the academic world. The case has been made with reference to five developments in particular:

N. Harris, *The Thirteenth-Century Animal Turn*,
https://doi.org/10.1007/978-3-030-50661-2_6

1. The availability in Latin, from the third decade of the thirteenth century, of an accessible translation of Aristotle's natural historical works, and the reception of these by philosopher-scientists such as Albert the Great, encylopaedists such as Thomas of Cantimpré and learned laymen such as Emperor Frederick II. The cumulative effect of the Aristotle-inspired writings on animals datable to between 1220 and 1270 was a highly significant one: they spread knowledge of long forgotten or previously unknown species, conveniently systematized what was known about the natural world and, perhaps most importantly in the long term, modelled and promoted an approach to looking at nature that relied more on observation and personal experience, and less on uncritical quotation from venerable authorities. Arguably for the first time since the classical era, animals came to be regarded as valuable and fascinating subjects of study in their own right.

2. An extension and revivification of the traditional use of animals to construct theological and moral meanings of relevance to human beings. This resulted from the – in some ways unlikely – joint influence of Aristotelian perspectives and new moral imperatives, felt in the wake of the Fourth Lateran Council of 1215, to communicate the doctrinal and ethical demands of Catholic Christianity to a broad-based lay audience. The ancient idea of nature as a book through which God reveals himself to people was not discredited; but it was significantly modified – with the result that it took on board new animals and new meanings, encompassed new literary forms, was restructured after the manner of a thirteenth-century nature encyclopaedia, and above all began to be revised on the basis of such classic Aristotelian priorities as observation and rational plausibility.

3. A marked increase in the frequency and creativity with which animals were used to construct aristocratic identity, during a troubled, unsettled period in the history of chivalry. Of particular importance here are heraldic emblems, a large proportion of which depicted animals, and the use of which in the thirteenth century went far beyond their original purpose as means of identification. Such emblems were widely used to connote both individual and dynastic identity; and they, and some of the animals they represented, were put to highly inventive use also in works of courtly literature that thematized issues concerning thirteenth-century knighthood.

4. A heightened tendency to challenge – as distinct from definitively refute – the age-old belief in humans' supremacy over, and qualitative difference from, the animal world. This questioning attitude can be seen in many contexts we have alluded to: in the rather uncomfortable and muddled treatment of the subject in some scientific literature; in works of imaginative fiction that posit notably porous, symbiotic relationships between knights and their horses; in a – for a time – almost modish fascination with the civilized, courtly centaur; and in a number of outlandish legal prosecutions of animals – based, as they seem to have been, on an instinctive assumption that their physical and moral capacities can barely be differentiated from those of humans.

5. Evidence of greater affection being felt, or at least shown, towards domestic companion animals – be it a knight's destrier or, most strikingly, a lady's lapdog. This development, however, existed in a state of (unconscious or hypocritical) tension with a continued use of unrestrained violence towards animals that retains the power to shock and disgust even averagely animal-friendly modern readers – and that seems indeed to have been exacerbated by some thirteenth-century people's physically aggressive reactions to the prejudice of their own innate superiority to animals being questioned or contested. This tension between violence and affection seems to have been as intrinsic to the thirteenth century as it is today, and to have characterized a wide range of contexts, ranging from comic tales to the life of St Francis of Assisi.

II

At the end we must essay a (very) brief survey of the extent to which these developments continued to be influential beyond the thirteenth century. The scientific progress made by the thirteenth-century followers of Aristotle seems generally to have satisfied their fourteenth- and fifteenth-century successors, but not those who, in the early modern period, sought to expand knowledge by means of observing – and also collecting – natural historical phenomena, and subjecting them to ever more stringent classification. With reference to England, for example, Keith Thomas (52) speaks of "an unbroken succession of active field naturalists, from William Turner (born 1508) to John Ray (born 1705)", who were in turn "members of a wider European scientific community" – a community which was to include such luminaries as Ulisse Aldrovandi, Conrad Gessner and the Comte de Buffon. Such figures also, however, very much saw themselves as Aristotle's disciples; and it would not be an exaggeration to claim that the scholarly methods and habits he inaugurated and thirteenth-century scholars began to apply retained a significant level of authority well into the modern period.

The use of symbolic or allegorical animals for the purposes of theological and moral instruction also continued apace in the later Middle Ages and beyond. Indeed, several of the trends we observed in our third chapter (the use of "new" animals, the introduction of new spiritual and secular meanings, the tendency towards systematization, the grounding of symbolism in "real-life" natural characteristics) were if anything still more pervasive and influential in the fourteenth and fifteenth centuries than they had been in the thirteenth – as indeed was the popularity of preaching. A certain caesura was to come, of course, with the Reformation, in the light of which animal imagery became at once more confessional, more political and more visual; and the two latter categories at least apply equally to the emblem books of the later sixteenth and seventeenth centuries. These tended also to be much more heavily indebted to classical sources and ethics than their medieval predecessors; and all in all one could not really claim that the medieval model of the Book of Nature survived the Baroque period intact. That said, some individual images at least were to prove remarkably long-lived: even today we might spontaneously refer to someone being eagle-eyed, having a basilisk stare, or rising like a phoenix from the ashes, without pausing to reflect that we are, in effect, quoting from the *Physiologus*.

An earlier casualty than the spiritual interpretation of animals was the use of animals to construct chivalric identity. As we have seen, the thirteenth century was remarkably productive in this regard, but already in the fourteenth century the whole edifice of chivalry became embroiled in what was to prove a terminal crisis, as "the devastation, misery and social dislocation caused by warfare focused ... general and widespread attention on the problem of the violence of the knightly order, and so on the relation between the ideals of chivalry and the value of peace" (Keen 1996, 9). That said, of course, heraldic emblems continued to be used in tournaments and other forms of post-courtly display, became more and more rigidly standardized and codified, and remain in use today in an astonishingly wide range of civic, sporting and political contacts – in reunified Germany, for example, no fewer than fifteen of the sixteen *Bundesländer* into which the country is now divided sport a heraldic animal or animals on their coat of arms (the sole exception being Saxony).

Meanwhile the late twelfth and thirteenth centuries' penchant for questioning and challenging conventional ideas of the anthropological difference between human and animals bore little perceptible fruit until the sixteenth century. Then, Michel de Montaigne's famous question, "when I am playing with my cat, who knows whether she is amusing herself with me more than I am with her?" provocatively raised the possibility that animals at least *might* possess cognitive and communicative abilities that approach, or even exceed, those of humans; and it led to problems with the orthodox view of human superiority being extensively discussed by leading philosophers such as Descartes, Locke, Hume and Bentham. The truly revolutionary thinker on these matters was, however, Darwin, since the publication of whose *Origin of Species* in 1859 it has been extremely difficult to deny that we humans are ourselves animals, and that the many characteristics and abilities we possess are therefore by definition animal ones.

If the tension between humanity and animality with which our medieval authors struggled has – at least in theory – long been resolved, the tension between treating non-human animals with affection and violence remains a contemporary, indeed perhaps a timeless one. Much progress has, of course, been made: out and out cruelty to animals is nowadays widely proscribed (with whatever success) by law; most people believe that animals have certain basic rights which should be protected; and our understanding of and respect for animal intelligence is growing all the time. Yet many issues remain. At the time of writing, indeed, several of these are being highlighted by the Covid-19 virus, whose causes and consequences clearly have much to do with the way human beings continue to mistreat animals. The "spillover" of animal diseases on to humans occasioned by the exploitation of wildlife and its concomitant loss of habitat; the deleterious results of an excessive reliance on antibiotics in intensive meat farming; the potentially uncontrollable effects on health of "wet" live animal markets – all these and other factors have been plausibly adduced as in part responsible for the calamity which befell much of the world in early 2020. Perhaps one day we will learn that, in order for human beings to thrive, animals must be allowed to do so as well. But that is a hard lesson, and one that will not be learnt by all overnight.

BIBLIOGRAPHY

This bibliography is in two parts. The first is a list of primary texts from the classical period and the Middle Ages; the second gives details of scholarly works from the twentieth and twenty-first centuries. English translations which are quoted from in the main text are marked with an asterisk.

PRIMARY SOURCES

Alan of Lille (Alanus ab Insulis): *De fide catholica contra haereticos,* in Jacques-Paul Migne (ed.), *Patrologiae cursus completus. Series Latina,* 221 vols, Paris 1844–64, vol. 210, cols 305–430

Albertus Magnus: *De animalibus libri XXVI,* ed. Hermann Stadler, 2 vols (Beiträge zur Geschichte der Philosophie und Theologie des Mittelalters 15–16), Münster: Aschendorff, 1916–20. Translation: Kenneth F. Kitchell and Irven M. Resnick, *Albertus Magnus On Animals: A Medieval Summa Theologica,* Baltimore/London: Johns Hopkins University Press, 1999*

Alexander von Roes: *Schriften,* ed. Herbert Grundmann and Hermann Heimpel (Deutsches Mittelalter. Kritische Studientexte der Monumenta Germaniae Historica 4), Weimar: Böhlau, 1949

Alexandre de Paris: *The Medieval French 'Roman d'Alexandre'. Vol. II: Version of Alexandre de Paris,* ed. E. C. Armstrong et al. (Elliott Monographs in the Romance Languages and Literatures 37), Princeton: Princeton University Press, 1937

Aristotle: *Historia animalium,* ed. and trans. A. L. Peck and D. M. Balme, 3 vols (Loeb Classical Library 437–9), Cambridge, MA: Harvard University Press, 1965–91

© The Author(s) 2020
N. Harris, *The Thirteenth-Century Animal Turn,*
https://doi.org/10.1007/978-3-030-50661-2

Idem: *Parts of Animals. Movement of Animals. Progression of Animals,* ed. and trans. A. L. Peck and E. S. Forster (Loeb Classical Library 323), Cambridge, MA: Harvard University Press, 1937

Bartholomaeus Anglicus: *De proprietatibus rerum,* Nuremberg: Koberger, 1492

Bataille d'Aliscans: La versione franco-italiana della 'Bataille d'Aliscans'. Codex Marcianus fr. VIII [=252], ed. Günter Holtus (Beihefte zur Zeitschrift für romanische Philologie 205), Tübingen: Niemeyer, 1985

Baudouin de Condé: *Dits et contes de Baudouin de Condé et de son fils Jean,* ed. Auguste Scheler, 3 vols, Brussels: Devaux, 1866–7

Berchorius, Petrus (Pierre Bersuire): *Reductorium morale,* Venice: Scotus, 1583

Berthold von Regensburg: *Vollständige Ausgabe seiner Predigten,* ed. Franz Pfeiffer and Joseph Strobl, 2 vols, Vienna: Braumüller, 1862–80

Beues of Hamtoun: The Romance of 'Sir Beues of Hamtoun', ed. Eugen Kölbing, 3 vols (Early English Text Society Extra Series 46, 48, 65), London: Kegan Paul, 1885–94

'Boeve de Haumtone' and 'Gui de Warewic'. Two Anglo-Norman Romances, trans. Judith Weiss (Medieval and Renaissance Texts and Studies 332), Tempe: Arizona Center for Medieval and Renaissance Studies, 2008

Brunetto Latini: *Li Livres du tresour,* ed. Francis J. Carmody, 2 vols (University of California Publications in Modern Philology 22), Berkeley: University of California Press, 1948

Caesarius von Heisterbach: *Dialogus Miraculorum – Dialog über die Wunder,* ed. and trans. Nikolaus Nösges and Horst Schneider, 5 vols (Fontes Christiani 86), Turnhout: Brepols, 2009

'La Chastelaine de Vergi'. Conte du XIIIe siècle, ed. and trans. Joseph Bédier, Paris: Piazza, 1927

Chaucer, Geoffrey: *The Complete Works,* ed. F. N. Robinson, 2nd edition, Oxford: Oxford University Press, 1974

Dante Alighieri: *La Divina Commedia,* ed. Umberto Bosco and Giovanni Reggio, 3 vols, Florence: Le Monnier, 1979

Idem: *De vulgari eloquentia,* ed. and trans. Steven Botterill, Cambridge: Cambridge University Press, 1996*

Etymachia: The Latin and German 'Etymachia', ed. Nigel Harris (Münchener Texte und Untersuchungen 102), Tübingen: Niemeyer, 1994

Frauenlob: *Leichs, Sangsprüche, Lieder,* ed. Karl Stackmann and Karl Bertau, 2 vols (Abhandlungen der Akademie der Wissenschaften, Philologisch-historische Klasse, 3rd series, 119–20), Göttingen: Vandenhoeck & Ruprecht, 1981

Friderici Romanorum Imperatoris Secundi 'De arte venandi cum avibus', ed. Carl Arnold Willemsen, Leipzig: Insel, 1942. Translation: *The Art of Falconry,* trans. Casey A. Wood and F. Marjorie Fyfe, Stanford: Stanford University Press, 1943*

Der Göttweiger Trojanerkrieg, ed. Alfred Koppitz (Deutsche Texte des Mittelalters 29), Berlin: Weidmann, 1926

Gottfried von Strassburg: *Tristan*, ed. Friedrich Ranke, Berlin: Weidmann, 1930. Translation: A. T. Hatto, Harmondsworth: Penguin, 1960*

Gritsch, Johannes: *Quadragesimale de tempore et de sanctis*, Lyon: Huguetan, 1506

Hartmann von Aue: *Iwein*, ed. Ludwig Wolff, 2 vols, Berlin/New York: de Gruyter, 1968. Translation: Frank Tobin et al., *The Complete Works of Hartmann von Aue*, University Park, PA: Pennsylvania State University Press, 2001, 235–321*

Heinrich von dem Türlin: *Diu Crône*, ed. G. H. F. Scholl (Bibliothek des Litterarischen Vereins in Stuttgart 27), Stuttgart: Litterarischer Verein, 1852

Henri de Valenciennes: *The Lay of Aristote*, ed. and trans. Leslie C. Brook and Glyn S. Burgess (Liverpool Online Series – Critical Editions of French Texts 16), Liverpool: School of Cultures, Languages and Area Studies, University of Liverpool, 2011*

Herbort von Fritzlar: *Liet von Troye*, ed. Georg Karl Frommann (Bibliothek der gesammten deutschen National-Litteratur 5), Quedlinburg/Leipzig: Basse, 1837

Holtnicker, Conradus (Ps. Bonaventura): *Sermones de tempore et de sanctis*, Paris: Badius Ascensius, 1521

Isidore of Seville (Isidor Hispalensis): *Etymologiarum sive originum libri XX*, ed. W. M. Lindsay, 2 vols, Oxford: Oxford University Press, 1911

Jacques de Vitry: *Histoire orientale – Historia orientalis*, ed. and trans. Jean Donnadieu, Turnhout: Brepols, 2008

Idem: *The Exempla or Illustrative Stories from the Sermons of Jacques de Vitry*, ed. Thomas Frederick Crane, London: Nutt, 1890

Jansen Enikel: *Werke*, ed. Philipp Strauch (Monumenta Germaniae Historica, Deutsche Chroniken 3), Hanover: Hahn, 1900

John of San Gimignano (Johannes Gorus): *Summa de exemplis et similitudinibus rerum locupletissima*, Lyon: Beraud and Michael, 1585

Konrad von Megenberg: *'Das Buch der Natur'. Kritischer Text nach den Handschriften*, ed. Robert Luff and Georg Steer (Texte und Textgeschichte 54), Tübingen: Niemeyer, 2003

Konrad von Stoffeln: *Gauriel von Muntabel*, ed. and trans. Siegfried Christoph (Arthurian Archives 15), Cambridge: Brewer, 2007*

Konrad von Würzburg: *Die Goldene Schmiede*, ed. Wilhelm Grimm, Berlin: Klemann, 1840

Idem: *Kleinere Dichtungen*, ed. Edward Schröder, 3 vols, Berlin: Weidmann, 1926

Idem: *Der Trojanische Krieg*, ed. Adelbert von Keller (Bibliothek des Litterarischen Vereins in Stuttgart 44), Stuttgart: Litterarischer Verein, 1858

Marcus of Orvieto: *Marci de Urbe Veteri, O.F.M., 'Liber de moralitatibus'*, ed. Girard J. Etzkorn, 3 vols, St. Bonaventure, NY: St. Bonaventure University, 2005

Der Meißner der Jenaer Liederhandschrift, ed. Georg Objartel (Philologische Studien und Quellen 85), Berlin: Schmidt, 1977

Der junge Meißner: Sangsprüche, Minnelieder, Meisterlieder, ed. Günter Peperkorn (Münchener Texte und Untersuchungen 79), Munich: Artemis, 1982

Der Münchner Oswald, ed. Michael Curschmann (Altdeutsche Textbibliothek 76), Tübingen: Niemeyer, 1974

Neckam, Alexander: *De naturis rerum*, ed. Thomas Wright (Rolls Series 34), London: Longman, 1863

Das Nibelungenlied, ed. Karl Bartsch and Helmut de Boor, Wiesbaden: Brockhaus, 1956. Translation: Cyril Edwards, Oxford: Oxford University Press, 2010*

Nicole de Margival: *Le Dit de la panthère*, ed. Bernard Ribémont (Classiques Français du Moyen Age 136), Paris: Champion, 2000

Novellistik des Mittelalters: Märendichtung, ed. Klaus Grubmüller (Bibliothek des Mittelalters 23), Frankfurt: Deutscher Klassiker Verlag, 1996

Pliny (Gaius Plinius Secundus): *Naturalis historiae libri XXXVII*, ed. Ludwig von Jan, 6 vols, Leipzig: Teubner, 1865–80

Polo, Marco: *The Travels*, trans. William Marsden, Ware: Wordsworth, 1997*

Richard de Fournival: *Le Bestiaire d'amour*, in Gabriel Bianciotto (trans.), *Bestiaires du Moyen Age*, Paris: Stock, 1980, 108–45

Rudolf von Ems: '*Alexander*'. *Ein höfischer Versroman des 13. Jahrhunderts*, ed. Victor Junk, 2 vols (Bibliothek des Litterarischen Vereins in Stuttgart 272, 274), Leipzig: Hiersemann, 1928–9

Seuse, Heinrich: *Deutsche Schriften*, ed. Karl Bihlmeyer, Stuttgart: Metzler, 1907

Sibote: *Frauenzucht*, in Karl Lambel (ed.), *Erzählungen und Schwänke* (Deutsche Classiker des Mittelalters 12), Leipzig: Brockhaus, 1883, 323–48

Soccus (Konrad von Brundelsheim): *Sermones de tempore et de sanctis*, 2 vols, Deventer: Pafraet, 1480

Solinus, Caius Julius: *Collectanea rerum memorabilium*, ed. Theodor Mommsen, Berlin: Weidmann, 1895

Stephen of Bourbon: *Stephani de Borbone 'Tractatus de diversis materiis predica-bilibus'*, ed. Jacques Berlioz et al., 2 vols to date (Corpus Christianorum. Continuatio Mediaevalis 124–124A), Turnhout: Brepols, 2005 and 2015

Thomas of Cantimpré (Thomas Cantimpratensis): *Liber de natura rerum*, ed. Helmut Boese, Berlin/New York: de Gruyter, 1973

Thomas of Celano: '*Vita Prima*', in *Analecta Francescana* 10 (1941), 4–115

Thomas of Chobham: *Summa de arte predicandi*, ed. Franco Morenzoni (Corpus Christianorum. Continuatio Mediaevalis 82), Turnhout: Brepols, 1988

Tuscan Poetry of the 'Duecento': An Anthology, ed. and trans. Frede Jensen (Garland Library of Medieval Literature 99), New York: Garland, 1994*

Ulrich von Etzenbach: *Alexander*, ed. Wendelin Toischer (Bibliothek des Litterarischen Vereins in Stuttgart 183), Tübingen: Fues, 1888

Vincent of Beauvais (Vincentius Bellovacensis): *Speculum naturalis* (= *Speculum quadruplex sive maius*, vol. 1), Douai: Beller 1624

P. Virgilii Maronis Opera, ed. R. A. B. Mynors, Oxford: Clarendon Press, 1969

Walther von der Vogelweide: *Werke. Band 1: Spruchlyrik*, ed. Günther Schweikle (Universal-Bibliothek 819), Stuttgart: Reclam, 1994
Wigamur, ed. Nathanael Busch, Berlin/New York: de Gruyter, 2009
William of Tyre: *A History of Deeds Done Beyond the Sea*, trans. Emily Atwater Babcock and A. C. Krey, 2 vols, New York: Columbia University Press, 1943
Wirnt von Grafenberg: *Wigalois,* ed. Sabine Seelbach and Ulrich Seelbach, Berlin/ New York: de Gruyter, 2014
Wolfram von Eschenbach: *Parzival*, ed. Karl Lachmann, 6th edition, Berlin: de Gruyter, 1998. Translation: A. T. Hatto, Harmondsworth: Penguin, 1980*
Idem: *Willehalm*, ed. Joachim Heinzle (Altdeutsche Textbibliothek 108), Tübingen: Niemeyer, 1994. Translation: Marion E. Gibbs and Sidney M. Johnson, Harmondsworth: Penguin, 1984*

CRITICAL LITERATURE

Anzulewicz, Henryk: 'Albertus Magnus und die Tiere', in Obermaier 2009 (see below), 29–54
Idem: 'Anthropology: The Concept of Man in Albert the Great', in Resnick (see below), 325–46
Bachmann-Medick, Doris: *Cultural Turns. New Orientations in the Study of Culture,* Berlin/New York: de Gruyter, 2016
Badel, Pierre-Yves: *Le 'Roman de la Rose' au XIVe siècle. Etude de la réception de l'oeuvre* (Publications romanes et françaises 153), Geneva: Droz, 1980
Barnes, Jonathan: *Aristotle: A Very Short Introduction,* Oxford: Oxford University Press, 2000
Barton, Ulrich: '*manheit* und *minne*. Achills zweifache Erziehung bei Konrad von Würzburg', in Henrike Laehnemann and Sandra Linden (eds), *Dichtung und Didaxe. Lehrhaftes Sprechen in der deutschen Literatur des Mittelalters*, Berlin/ New York: de Gruyter, 2009, 189–204
Berlioz, Jacques, and Marie-Anne Polo de Beaulieu: 'Les Recueils d'exempla et la diffusion de l'encyclopédisme médiéval', in Picone (see below), 179–212
Bichon, Jean: *L'Animal dans la littérature française au XIIe et au XIIIe siècle*, 2 vols, PhD dissertation, Université de Lille, 1973
Binkley, Peter: 'Preachers' Responses to Thirteenth-Century Encyclopaedism', in idem (ed.), *Pre-Modern Encyclopaedic Texts: Proceedings of the Second COMERS Congress, Groningen, 1–4 July 1996* (Brill's Studies in Intellectual History 79), Leiden: Brill, 1997, 75–88
Blaschitz, Gertrud: 'Die Katze', in eadem et al. (eds), *Symbole des Alltags – Alltag der Symbole. Festschrift für Harry Kühnel zum 65. Geburtstag*, Graz: Akademische Drucks- und Verlagsanstalt, 1992, 589–616
Borgards, Roland: 'Introduction: Cultural and Literary Animal Studies', in *Journal of Literary Theory* 9 (2015), 155–60

Idem (ed.): *Tiere. Kulturwissenschaftliches Handbuch*, Stuttgart: Metzler, 2016

Bowden, Sarah: *Bridal-Quest Epics in Medieval Germany: A Revisionary Approach*, London: Modern Humanities Research Association, 2012

Brackert, Helmut: '*deist rehtiu jegerie*. Höfische Jagddarstellungen in der deutschen Epik des Mittelalters', in Rösener (see below), 365–406

Camille, Michael: 'Bestiary or Biology? Aristotle's Animals in Oxford, Merton College, Ms 271', in C. Steel (see below), 355–96

Classen, Albrecht: 'The Epistemological Function of Monsters in the Middle Ages', in *Lo Sguardo* 9 (2012), 13–34

Cohen, Jeffrey J.: *Medieval Identity Machines* (Medieval Cultures 35), Minneapolis: University of Minnesota Press, 2003

Crane, Susan: *Animal Encounters. Contacts and Concepts in Medieval Britain*, Philadelphia: University of Pennsylvania Press, 2012

Delort, Robert: *Les Animaux ont une histoire*, Paris: Seuil, 1984

DeMello, Margo: *Animals and Society. An Introduction to Human-Animal Studies*, New York: Columbia University Press, 2012

Descola, Philippe: *Par-delà nature et culture*, Paris: Gallimard, 2005

Dinzelbacher, Peter: *Mensch und Tier in der Geschichte Europas*, Stuttgart: Kröner, 2000

Idem: 'Animal Trials: A Multidisciplinary Approach', in *Journal of Interdisciplinary History* 32 (2002), 405–21

Dowling, Abigail P.: 'Landscape of Luxuries: Mahaut d'Artois's (1302–1329) Management and Use of the Park at Hesdin', in Albrecht Classen (ed.), *Rural Space in the Middle Ages and Early Modern Age: The Spatial Turn in Premodern Studies* (Fundamentals of Medieval and Early Modern Culture 9), Berlin/ Boston: de Gruyter, 2012, 367–88

Einhorn, Jürgen Werinhard: *Spiritalis Unicornis: Das Einhorn als Bedeutungsträger in Literatur und Kunst des Mittelalters* (Münstersche Mittelalter-Schriften 13), Munich: Fink, 1976

Encyclopaedia Britannica, https://www.britannica.com/biography/Saint-Albertus-Magnus, accessed 6th April 2020

Ertzdorff, Xenja von (ed.): *Die Romane von dem Ritter mit dem Löwen* (Chloe 20), Amsterdam/Atlanta: Rodopi, 1994

Etzkorn, Girard J.: 'Marcus of Orvieto's *Liber de moralitatibus*', in *Mediaevalia. Textos e Estudos* 23 (2004), 187–92

Fried, Johannes: 'Kaiser Friedrich II als Jäger', in Rösener (see below), 1997, 149–66

Friedman, John B.: 'Albert the Great's Topoi of Direct Observation and His Debt to Thomas of Cantimpré', in Binkley (see above), 1997, 379–92

Friedrich, Udo: *Menschentier und Tiermensch. Diskurse der Grenzziehung und Grenzüberschreitung im Mittelalter* (Historische Semantik 5), Göttingen: Vandenhoeck & Ruprecht, 2009

Fudge, Erica: *Animal*, London: Reaktion, 2002

Gerhardt, Christoph: *Die Metamorphosen des Pelikans. Exempel und Auslegung in der mittelalterlichen Literatur* (Trierer Studien zur Literatur 1), Frankfurt: Lang, 1979

Giese, Martina: 'Die Tierhaltung am Hof Kaiser Friedrichs II. zwischen Tradition und Innovation', in Knut Görich et al. (eds), *Herrschaftsräume, Herrschaftspraxis und Kommunikation zur Zeit Kaiser Friedrichs II.*, Munich: Utz, 2008, 121–71

Glock, Hans-Johann: 'Philosophie – Geist der Tiere', in Borgards 2016 (see above), 60–78

Gottschall, Dagmar: 'Albert's Conributions to and Influence on Vernacular Literatures', in Resnick (see below), 725–57

Guerrini, Anita: 'Animals and Ecological Science', in Kalof (see below), 489–505

Harris, Nigel: *The Latin and German 'Etymachia'. Textual History, Edition, Commentary* (Münchener Texte und Untersuchungen 102), Tübingen: Niemeyer, 1994

Idem: 'Willehalm and Puzzât, Guillelme and Baucent: The Hero and His Horse in Wolfram's *Willehalm* and in the *Bataille d'Aliscans'*, in Michael Butler and Robert Evans (ed.), *The Challenge of German Culture. Essays Presented to Wilfried van der Will*, Basingstoke: Routledge, 2000, 13–24

Idem: 'Animal, Vegetable, Mineral. Some Observations on the Presentation and Function of Natural Phenomena in *Willehalm* and in the Old French *Aliscans'*, in Jones/McFarland (see below), 2002, 211–29

Idem: 'The Camel: Perspectives and Meanings in Medieval Literature', in S. Hartmann (see below), 2007, 113–31

Idem: 'Der Pfau bei Konrad von Megenberg – und anderswo', in Edith Feistner (ed.), *Konrad von Megenberg (1309–1374): Ein spätmittelalterlicher 'Enzyklopädist' in europäischem Kontext* (*Jahrbuch der Oswald von Wolkenstein-Gesellschaft* 18), Wiesbaden: Reichert, 2011, 175–88

Hartmann, Heiko: 'Tiere in der historischen und literarischen Heraldik des Mittelalters. Ein Aufriss', in Obermaier 2009 (see below), 147–79

Hartmann, Sieglinde (ed.): *Fauna and Flora in the Middle Ages. Studies of the Medieval Environment and its Impact on the Human Mind* (Beihefte zur Mediaevistik 8), Frankfurt: Lang, 2007

Haug, Walter: 'Das Komische und das Heilige: Zur Komik in der religiösen Literatur des Mittelalters', in *Wolfram-Studien* 7 (1982), 8–31

Hayward, Jane: 'Sacred Vestments as they Developed in the Middle Ages', in John T. Doherty et al. (eds), *Ecclesiastical Vestments of the Middle Ages: An Exhibition* (*The Metropolitan Museum of Art Bulletin* 29/7, 1971), 299–310

Henkel, Nikolaus: *Studien zum 'Physiologus' im Mittelalter* (Hermaea N. S. 38), Tübingen: Niemeyer, 1976

Hills, Richard L.: 'Early Italian Papermaking, a Crucial Technical Revolution', in *IPH Yearbook* 9 (1992), 37–46

Hoogvliet, Margriet: 'Mappae mundi and Medieval Encyclopaedias', in Binkley (see above), 63–74

Hüntelmann, Axel: 'Institutionen und Praktiken: Geschichte des Tierversuchs', in Borgards 2016 (see above), 160–73

Jackson, Christine E.: *Peacock*, London: Reaktion, 2006

Jackson, Deirdre: *Lion*, London: Reaktion, 2010

Jones, Malcolm H.: 'Cats and Cat-skinning in Late Medieval Art and Life', in S. Hartmann (see above), 97–112

Jones, Martin H., and Timothy McFarland (eds), *Wolfram's 'Willehalm': Fifteen Essays*, Columbia, SC: Camden House, 2002

Kaeuper, Richard W.: *Medieval Chivalry*, Cambridge: Cambridge University Press, 2016

Kalof, Linda (ed.): *The Oxford Handbook of Animal Studies*, Oxford/New York: Oxford University Press, 2017

Kay, Sarah: *Animal Skins and the Reading Self in Medieval Latin and French Bestiaries*, Chicago: University of Chicago Press, 2017

Keen, Maurice: *Chivalry*, New Haven/London: Yale University Press, 1984

Idem: *Nobles, Knights and Men-at-Arms in the Middle Ages*, London: Bloomsbury, 1996

Kern, Manfred, and Alfred Ebenbauer: *Lexikon der antiken Gestalten in den deutschen Texten des Mittelalters*, Berlin/New York: de Gruyter, 2003

Kompatscher, Gabriela et al. (eds): *Human-Animal Studies: Eine Einführung für Studierende und Lehrende* (UTB 4759), Münster/New York: Waxmann, 2017

Kraß, Andreas: 'Die Spur der Zentauren. Pferde- und Eselsmänner in der deutschen Literatur des Mittelalters', in Hans Jürgen Scheuer and Ulrike Vedder (eds), *Tier im Text. Exemplarität und Allegorizität literarischer Lebewesen* (Publikationen zur *Zeitschrift für Germanistik* 29), Berne: Lang, 2015, 81–96

Lecouteux, Claude: *Les Monstres dans la littérature allemande du moyen âge. Contribution à l'étude du merveilleux médiéval* (Göppinger Arbeiten zur Germanistik 330), 3 vols, Göppingen: Kümmerle, 1982

Lidaka, Juris G.: 'Bartholomaeus Anglicus in the Thirteenth Century', in Binkley (see above), 393–406

Lim, Gary: '"A Stede Gode and Lel": Valuing Arondel in *Bevis of Hampton*', in *Postmedieval* 2 (2011), 50–68

Lohr, Charles H.: 'Medieval Latin Aristotle Commentaries. Authors A–F', in *Traditio* 23 (1967), 313–413

Lother, Helmut: *Der Pfau in der altchristlichen Kunst: Eine Studie über das Verhältnis von Ornament und Symbol*, Leipzig: Dieterich, 1929

McCracken, Peggy: *In the Skin of a Beast: Sovereignty and Animality in Medieval France*, Chicago: University of Chicago Press, 2017

Meier, Christel: 'Organisation of Knowledge and Encyclopaedic *Ordo*: Functions and Purposes of a Universal Literary Genre', in Binkley (see above), 103–26

Meyer, Heinz: *Die Enzyklopädie des Bartholomäus Anglicus: Untersuchungen zur Überlieferungs- und Rezeptionsgeschichte von 'De proprietatibus rerum'* (Münstersche Mittelalter-Schriften 77), Munich: Fink, 2000

Moore, John C.: *Pope Innocent III (1160/61–1216): To Root Up and to Plant,* Leiden: Brill, 2003

Morris, Desmond: *Leopard,* London: Reaktion, 2014

Mowatt, D. G., and Hugh Sacker: *'The Nibelungenlied'. An Interpretative Commentary,* Toronto: University of Toronto Press, 1967

Müller, Lothar: *Weiße Magie. Die Epoche des Papiers,* Munich: Hanser, 2012

Obermaier, Sabine: 'Löwe, Adler, Bock. Das Tierrittermotiv und seine Verwandlungen im späthöfischen Artusroman', in Bernhard Jahn and Otto Neudeck (eds), *Tierepik und Tierallegorese. Studien zur Poetologie und historischen Anthropologie vormoderner Literatur* (Mikrokosmos 71), Frankfurt: Lang, 2004, 121–39

Eadem (ed.): *Tiere und Fabelwesen im Mittelalter,* Berlin/New York: de Gruyter, 2009

Oeser, Erhard: *Katze und Mensch. Die Geschichte einer Beziehung,* Darmstadt: Wissenschaftliche Buchgesellschaft, 2005

Oggins, Robin S.: 'Falconry and Medieval Social Status', in *Mediaevalia* 12 (1989), 43–55

Ohly, Friedrich: 'The Spiritual Sense of Words in the Middle Ages', trans. David A. Wells, in *Forum for Modern Language Studies* 41 (2005), 18–42

Pastoureau, Michel: *L'Ours. Histoire d'un roi déchu,* Paris: Seuil, 2007

Peters, Edward: *The Magician, the Witch, and the Law,* Philadelphia: University of Pennsylvania Press, 1978

Picone, Michelangelo (ed.): *L'enciclopedismo medievale,* Ravenna: Longo, 1994

Resnick, Irven M. (ed.): *A Companion to Albert the Great. Theology, Philosophy, and the Sciences* (Brill's Companions to the Christian Tradition 38), Leiden: Brill, 2013

Ribémont, Bernard: 'Code courtois ou morale religieuse? Le jeu sur une ambiguité: l'exemple de Nicole de Margival et de Jean de Condé', in *Revue des Langues Romanes* 109 (2005), 199–212

Ritvo, Harriet: 'On the Animal Turn', in *Daedalus* 136/4 (2007), 118–22

Rösener, Werner (ed.): *Jagd und höfische Kultur im Mittelalter,* Göttingen: Vandenhoeck & Ruprecht, 1997

Rogers, Katharine M.: *Cat,* London: Reaktion, 2006

Salisbury, Joyce E.: *The Beast Within. Animals in the Middle Ages,* 2nd edition, London/New York: Routledge, 2011

Salter, David: *Holy and Noble Beasts. Encounters with Animals in Medieval Literature,* Cambridge: Brewer, 2001

Saurma-Jeltsch, Lieselotte E.: 'Bucephalus als "Alter Ego" Alexanders des Großen', in Judith Klinger and Andreas Kraß (eds), *Tiere: Begleiter des Menschen in der Literatur des Mittelalters,* Cologne: Böhlau, 2017, 33–45

Scheibelreiter, Georg: *Wappen im Mittelalter,* Darmstadt: Primus, 2014

Schinagl, Elisabeth: *Naturkunde-Exempla in lateinischen Predigtsammlungen des 13. und 14. Jahrhunderts* (Lateinische Sprache und Literatur des Mittelalters 32), Berne: Lang, 2001

Schlögel, Karl: 'Kartenlesen, Augenarbeit: Über die Fälligkeit des "spatial turn" in den Geschichts- und Kulturwissenschaften', in Heinz Dieter Kittsteiner (ed.), *Was sind Kulturwissenschaften? 13 Antworten,* Munich: Fink, 2004, 261–83

Schmidtke, Dietrich: *Geistliche Tierinterpretation in der deutschsprachigen Literatur des Mittelalters, 1100–1500,* 2 vols, PhD dissertation, Freie Universität Berlin, 1966

Schumacher, Meinolf: *Ärzte mit der Zunge. Leckende Hunde in der europäischen Literatur* (Aisthesis Essay 16), Bielefeld: Aisthesis, 2003

Schwabe, Julius: 'Lebenswasser und Pfau. Zwei Symbole der Wiedergeburt', in *Symbolon* 1 (1960), 138–72

Signori, Gabriela: *Das 13. Jahrhundert: Eine Einführung in die Geschichte des spätmittelalterlichen Europas,* Stuttgart: Kohlhammer, 2007

Sorrell, Roger D.: *St. Francis of Assisi and Nature. Tradition and Innovation in Western Christian Attitudes toward the Environment,* New York/Oxford: Oxford University Press, 1988

Steel, Carlos et al. (eds): *Aristotle's Animals in the Middle Ages and Renaissance* (Mediaevalia Lovaniensia I/27), Leuven: Leuven University Press, 1999

Steel, Karl: *How to Make a Human: Animals and Violence in the Middle Ages,* Columbus: Ohio State University Press, 2011

Tesnière, Marie-Hélène: 'Le *Reductorium morale* de Pierre Bersuire', in Picone (see above), 229–49

Thomas, Keith: *Man and the Natural World: Changing Attitudes in England 1500–1800,* Harmondsworth: Penguin, 1984

Thompson, Augustine, O. P.: *Francis of Assisi. The Life,* Ithaca/London: Cornell University Press, 2013

Tillmann, Helene: *Papst Innocenz III.,* Bonn: Röhrscheid, 1954

Toepfer, Georg: 'Institutionen und Praktiken: Geschichte der Zoologie', in Borgards 2016 (see above), 139–48

van den Abeele, Baudouin: *La Fauconnerie dans les lettres françaises du XIIe au XIVe siècle* (Mediaevalia Lovaniensia I/18), Leuven: Leuven University Press, 1990

Idem: *La Fauconnerie au moyen âge. Connaissance, affaitage et médecine des oiseaux de chasse d'après les traités latins,* Paris: Klincksieck, 1994

Idem: 'Le *De animalibus* d'Aristote dans le monde latin: modalités de sa réception médiévale', in *Frühmittelalterliche Studien* 33, 287–318 (= van den Abeele 1999a)

Idem: 'L'Allégorie animale dans les encyclopédies latines du Moyen Âge', in Jacques Berlioz and Marie-Anne Polo de Beaulieu (eds), *L'Animal exemplaire au Moyen Âge (V^e–XVe siècles)*, Rennes: Presses Universitaires de Rennes, 1999, 123–43 (= van den Abeele 1999b)

Idem: 'Diffusion et avatars d'une encyclopédie: le *Liber de natura rerum* de Thomas de Cantimpré', in idem and Godefroid de Callataÿ (eds), *'Une Lumière venue d'ailleurs'. Héritages et ouvertures dans les encyclopédies d'Orient et d'Occident au Moyen Âge* (Réminiscences 9), Turnhout: Brepols, 2008, 141–76

Volfing, Annette: '*Parzival* and *Willehalm*: Narrative Continuity?', in Jones/McFarland (see above), 45–59

Wailes, Stephen L.: 'The Crane, the Peacock, and the Reading of Walther von der Vogelweide 19,29', in *Modern Language Notes* 88 (1973), 947–55

Walker-Meikle, Kathleen: *Medieval Pets*, Cambridge: Boydell & Brewer, 2012

White, Lynn, Jr.: 'The Historical Roots of Our Ecologic Crisis', in *Science* 155 (1967), 1203–7

Wild, Markus: 'Philosophie: Anthropologische Differenz', in Borgards 2016 (see above), 47–59

Wirth, Karl-August: '*Imperator pedes papae deosculatur*. Ein Beitrag zur Bildkunde des 16. Jahrhunderts', in Hans Martin Freiherr von Erffa and Elisabeth Herget (eds), *Festschrift für Harald Keller zum 60. Geburtstag*, Darmstadt: Roether, 1963, 175–221

Yamamoto, Dorothy: *The Boundaries of the Human in Medieval English Literature*, Oxford/New York: Oxford University Press, 2000

INDEX

This index includes the names of people mentioned in the text (with the exception of those of twentieth- and twenty-first-century scholars); the titles of anonymous medieval works; and the names of animals, by species (including beings we now know to be mythical).

© The Author(s) 2020
N. Harris, *The Thirteenth-Century Animal Turn*,
https://doi.org/10.1007/978-3-030-50661-2